CAMBRIDGE TRACTS IN MATHEMATICS

70. *Applications of sieve methods to the theory of numbers*

C. HOOLEY, Sc.D.

Professor of Pure Mathematics at University College, Cardiff
Sometime Fellow of Corpus Christi College, Cambridge

Applications of sieve methods to the theory of numbers

CAMBRIDGE UNIVERSITY PRESS

CAMBRIDGE

LONDON·NEW YORK·MELBOURNE

Published by the Syndics of the Cambridge University Press
The Pitt Building, Trumpington Street, Cambridge CB2 1RP
Bentley House, 200 Euston Road, London NW1 2DB
32 East 57th Street, New York, NY 10022, USA
296 Beaconsfield Parade, Middle Park, Melbourne 3206, Australia

First published 1976

Printed in Great Britain at the
University Printing House, Cambridge
(Euan Phillips, University Printer)

Library of Congress Cataloguing in Publication Data
Hooley, C 1928–
Applications of sieve methods to the theory of numbers
(Cambridge tracts in mathematics; 70)
Bibliography: p. 119
1. Sieves (Mathematics) I. Title. II. Series.
QA241. H64   512′.7   75–27796
ISBN 0 521 20915 3

Contents

Contents   vii

Preface

This tract has its origin in an essay that gained the author an Adams Prize of the University of Cambridge in 1973. Parts of some Chapters have therefore been based on the author's previously published papers as follows: Chapter 2, [37]; Chapter 3, [36]; Chapter 5, [32]; Chapter 6, [39].

The author is grateful to Dr Greaves, Professor Halberstam, and Dr Smithies for their valuable comments during the preparation of the tract. He is also greatly indebted to Mrs M. E. Coles for her skilful production of the typescript. Finally, he would like to thank the Cambridge University Press for their unfailing courtesy and help.

Introduction

The modern sieve method was developed in the expectation that it might lead to a proof of Goldbach's conjecture and other similar important conjectures in the theory of numbers. Although it is now generally conceded that this hope is not likely to be entirely realized, the method has been conspicuously successful in establishing analogous theorems in which prime numbers are replaced by numbers with a bounded number of prime factors. Moreover, the achievements of the sieve method have been by no means confined to the area of study for which it was initially developed. It has, indeed, made an invaluable contribution to the theory of numbers generally because of its power to elicit auxiliary results that are needed in the proofs of many theorems. In particular, it is to the perspicacity of Erdös that we owe our present appreciation of the manifold uses to which sieve methods can be put.

Characteristically the sieve method in its applications is often used alongside other methods but is seldom integrated with them. On the other hand the scope of the sieve method can sometimes be usefully enlarged if other techniques and ideas are closely linked with it.

The field of application of the sieve method, even in its present form, is probably by no means exhausted. There are, for example, problems about powers of integers and about sparse sequences to which sieve methods have a relevance hitherto not fully appreciated.

In the present tract we consider some important problems with the intention of exposing the potentialities of the sieve method when it is allied with other methods or when it is directed in an unfamiliar direction. The choice of illustrative topics, which has been much influenced by the author's own researches on the subject, involves the introduction of the sieve method in many of its known aspects.

Each problem is discussed in a separate chapter, there being a preparatory chapter that develops sieve methods *ab initio* as far as is needed for the subsequent applications. Though the tract is in no way intended to be a systematic treatise, this initial chapter may well serve as a useful introduction to a deeper study of sieve methods.

Sieve methods apart, the treatment presupposes a wide but not deep background in the theory of numbers. Most of the material assumed is classical or easily accessible, the main exception being the estimate in Chapter 2 for the Kloosterman sum, which can if necessary be assumed by the reader without affecting his understanding of the rest of the account.

Many important uses of the sieve method fall outside the confines of this tract. In particular, there are the interesting and complicated applications to problems of the Goldbach type, of which a full account appears in the recent monograph by Halberstam and Richert [25].

In this tract the term sieve is used in the classical context of a sifting process. However, after the introduction of Linnik's large sieve, the term has tended to be applied to trigonometrical inequalities from which large sieve results can be derived. Also, by association the term is now often used in connection with other inequalities of a similar type from which sieve results as such do not necessarily flow. We should therefore explain that we make no use of such inequalities here, even though they have already played a significant rôle in the theory of numbers.

Notation

Owing to the nature of the tract no attempt has been made to secure absolute consistency in the use of notation. In the light of the following summary the meaning of the notation should in all cases be plain from the context in which it is used.

The letters a, d, h, k, l, m, n, q, r, δ, μ, ν are usually positive integers; p is a prime, as is sometimes q; s, t are usually positive integers save when $s = \sigma + it$ is a complex variable; x, y are usually real variables, frequently regarded as tending to infinity, save when they are indeterminates in a quadratic or cubic form; ξ is usually a real number.

The letters ϵ, η are arbitrarily small positive constants that are not necessarily the same at each occurrence; A, A_i are positive absolute constants, while $A(\alpha)$ is a positive constant that depends at most on the parameter α. The equation $f = O(|g|)$ always denotes an inequality of the form $|f| \leqslant A|g|$ save where it is clear from the context that the weaker inequality

$$|f| \leqslant A(\epsilon)|g|$$

is being used instead; furthermore these inequalities are deemed to hold for all values of the variables that are consistent with stated or implied conditions.

The following functions occur: $d(n)$ the number of divisors of n; $r(n)$ the number of representations of n as a sum of two squares; $\sigma_\gamma(n)$ the sum of the γ-th powers of the divisors of n; $\omega(n)$ the number of distinct prime factors of n; $\Omega(n)$ the total number of prime factors of n; $\mu(n)$ the Möbius function; $\phi(n)$ the Euler totient function; $\Lambda(n)$ von Mangoldt's function. The (positive) highest common factor and least common multiple of l and m are denoted by (l, m) and $[l, m]$, respectively. The least absolute distance of ξ from an integer is denoted by $\|\xi\|$; $[x]$ is the greatest integer not exceeding x.

As customary $\pi(x)$ is the number of primes not exceeding x, while $\pi(x; a, k)$ is the number of such primes p for which $p \equiv a$, mod k. Also we write

$$\vartheta(x) = \sum_{p \leqslant x} \log p,$$

the cognate function $\theta(x)$ being defined on page 56.

1. *Survey of sieve methods*

1 General principles

The introduction of the sieve method into the theory of numbers is traditionally associated with Eratosthenes, to whom is attributed the procedure for determining the prime numbers between $\sqrt{x}$ and x by striking out from the numbers not exceeding x all those that are multiples of primes not exceeding $\sqrt{x}$. Eratosthenes's method already possesses the distinguishing feature of the sieve method in that it is concerned with counting the number of elements in a set that do not possess certain assigned properties. Since, however, the sieve method is primarily an exclusion process, its applications are not necessarily restricted, even in its more recondite forms, to the sifting of sequences or, indeed, to the theory of numbers. It is therefore appropriate that our initial description of the subject should be formulated in a general context.

Let S be a finite set of objects and let A be a finite family† of properties $\alpha_1, \ldots, \alpha_n$. Then our purpose is to discuss means of estimating the number of elements in the set that have none of the properties in A. To this end we introduce the following notation, where sub-sets of $(1, \ldots, n)$ are denoted by enumerative symbols of the type $(\iota_1, \ldots, \iota_q)$ in which $\iota_1, \ldots, \iota_q$ are distinct: $N(i_1, \ldots, i_r)$ is the number of elements in S having all the properties α_i that are indexed by $(i_1, \ldots, i_r)$, where we write N, the number of elements in S, when $(i_1, \ldots, i_r)$ is the empty set; $N_{j_1, \ldots, j_s}$ is the number of elements in S having none of the properties α_j that are indexed by $(j_1, \ldots, j_s)$; if $(i_1, \ldots, i_r)$ and $(j_1, \ldots, j_s)$ are disjoint, then $N_{j_1, \ldots, j_s}(i_1, \ldots, i_r)$ is the number of elements in S which have all the properties $\alpha_{i_1}, \ldots, \alpha_{i_r}$ but none of the pro-

---

† i.e. the properties $\alpha_1, \ldots, \alpha_n$ are not necessarily distinct. In specific applications of the ensuing principles it is therefore unnecessary to observe the formality of noting or verifying that the given properties are logically inequivalent. Previous formulations of these methods appear to have neglected this logical nicety.

perties $\alpha_{j_1}, ..., \alpha_{j_s}$; finally $M(i_1, ..., i_r)$ is the number of elements in S which have at least one of the properties α_i indexed by $(i_1, ..., i_r)$ (where $M(i) = N(i)$).

The fundamental relation in sieve theory is then that

$$N_{1, ..., n} = \sum_{(j_1, ..., j_s)} (-1)^s N(j_1, ..., j_s), \tag{1}$$

where the summation is over all sub-sets $(j_1, ..., j_s)$ of $(1, ..., n)$ including the empty set. This is most easily proved by considering first the special case in which S consists of a single element, the case where S is null being both trivial and uninteresting. If this element has none of the properties, then the right-hand side of (1) is 1 as required. If, on the other hand, it has a positive number of the properties corresponding to precisely $r (\leqslant n)$ indices, then the right-hand side of (1) is

$$1 - r + \binom{r}{2} - ... = (1-1)^r = 0,$$

as is also required. The general case is then immediately inferred by superposition.

There have been some important direct applications of this relation to the theory of numbers. In particular, it leads easily to Legendre's formula

$$\sum_{l'} \mu(l') \left[\frac{x}{l'} \right] = \sum_{l'} \mu(l') \left\{ \frac{x}{l'} + O(1) \right\} = x \prod_{p \leqslant \xi} \left(1 - \frac{1}{p} \right) + O(\sum_{l'} 1)$$

for the number of positive integers not exceeding x that are not divisible by primes p not exceeding ξ, where l' denotes, generally, a square-free number (possibly 1) with no prime factors exceeding ξ. Yet, as is familiar, this formula cannot be used in its present form to calculate effectively the number of non-excluded elements in the sieve of Eratosthenes because the remainder term is not satisfactorily estimated when ξ is large.

The limited scope of Legendre's formula is associated with the fact that (1) suffers, generally, from the defect of having too many terms. Sometimes, however, it turns out that further progress is immediately possible, since it can happen that many of these terms vanish in the circumstances of particular applications. As a precursor of the type of situation we have in mind, we present Legendre's formula in the obvious alternative form

$$\sum_{l' \leqslant x} \mu(l') \left[\frac{x}{l'}\right] = x \sum_{l' \leqslant x} \frac{\mu(l')}{l'} - \sum_{l' \leqslant x} \mu(l') \left(\frac{x}{l'} - \left[\frac{x}{l'}\right]\right)$$

$$= x \sum_{l' \leqslant x} \frac{\mu(l')}{l'} + O(\sum_{l' \leqslant x} 1),$$

in which the remainder term depends on fewer constituents than previously when ξ is large. Using only the estimate furnished by the final form of this remainder term, we are then able to consider substantially larger values of ξ than before, even though the sieve of Eratosthenes itself is still beyond the scope of the formula unless we appeal to deep results that it would be the primary object of this method to circumvent.† In other cases these improvements sometimes lead to a complete solution of multiplicative problems, one such example being mentioned later in §3. Problems of the latter type, however, are relatively rare, and in most instances are anyway more appropriately the subject of other methods. It is, therefore, necessary to seek more workable forms of (1).

Some useful variants of the formula can be derived in the most elementary way. We have, for instance, that

$$N_{1, \dots, n} \leqslant N_{1, \dots, r} \tag{2}$$

for $r < n$, so that an upper bound for $N_{1, \dots, n}$ can be approached through a form of (1) with fewer terms. Furthermore, we have

$$N_{1, \dots, n} \geqslant N_{1, \dots, r} - M(r+1, \dots, n),$$

---

† In view of the history of Legendre's formula, there is perhaps some interest in seeing how far this analysis can be taken by means of methods that do not implicitly involve the use of the prime number theorem. While somewhat of a diversion from the main subject matter of the tract, it therefore seems worth mentioning that a significant result can be obtained for

$$\xi = x^{(c \log \log \log x)/\log \log x}$$

by methods that use little more than the Mertens formulae, c being a sufficiently small positive constant. Thus, contrary to what has been normally asserted in the literature, the use of Legendre's formula leads to the bound

$$\pi(x) = O\left(\frac{x \log \log x}{\log x \log \log \log x}\right),$$

which is in fact better than the usual bounds that have been derived by the simplest version of Brun's method (cf. next paragraph but one).

so that
$$N_{1,\ldots,n} = N_{1,\ldots,r} + O\{M(r+1,\ldots,n)\}. \tag{3}$$
As for the error term in this, we have
$$M(r+1,\ldots,n) \leqslant \sum_{r+1 \leqslant i \leqslant n} N(i). \tag{4}$$
We are thus led to a simple formula for $N_{1,\ldots,n}$ by combining (1), (3), and (4). In practice this often leads to satisfactory asymptotic formulae so long as $N(i)$ is small when i is large. We therefore name this process the *simple asymptotic sieve*, a terminology that has not hitherto been used. Simple though it is, its potentialities have not always been as fully appreciated as they deserved. It is worth remarking, in addition, that the upper bound implicit in (2) may be expected to forecast the actual value of $N_{1,\ldots,n}$ in circumstances where there are reasons to suppose that the right-hand side of (4) is small. This comment is particularly relevant to the problems on primitive roots and power-free numbers that are discussed in Chapters 3 and 4.

In many of the more interesting problems there are too many elements with the unwanted properties for this simple idea to serve. However, in his first important paper on the sieve method Brun [4] introduced a more sophisticated method of reducing the number of terms in (1) that need be considered. Brun's formulae are
$$N_{1,\ldots,n} \leqslant \sum_{\substack{(j_1,\ldots,j_s) \\ s \leqslant k}} (-1)^s N(j_1,\ldots,j_s), \tag{5A}$$
if k is a given *even* integer not exceeding n, and
$$N_{1,\ldots,n} \geqslant \sum_{\substack{(j_1,\ldots,j_s) \\ s \leqslant k}} (-1)^s N(j_1,\ldots,j_s), \tag{5B}$$
if k is a given *odd* integer not exceeding n. In combination they give
$$N_{1,\ldots,n} = \sum_{\substack{(j_1,\ldots,j_s) \\ s < k}} (-1)^s N(j_1,\ldots,j_s) + O\Big(\sum_{(j_1,\ldots,j_k)} N(j_1,\ldots,j_k)\Big) \tag{6}$$
for k of either parity. These are easily verified by a modification of the method that was used for (1). Confining ourselves for brevity to (5A), we again consider the case where S is a single element. If this element has none of the properties, then the formula stands as before. If the element has a positive number

of the properties corresponding to precisely r indices, then the right-hand side of (5 A) is now

$$1 - r + \binom{r}{2} - \ldots + \binom{r}{k}, \tag{7}$$

which is the sum of the first $(k+1)$ terms in the binomial expansion of $(1-1)^r = 0$. If $k \leqslant \frac{1}{2}r$, the terms are in magnitude monotonically non-decreasing so that (7) is non-negative. If $k > \frac{1}{2}r$, then (7) is equal to

$$\binom{r}{k+1} - \binom{r}{k+2} + \ldots + (-1)^{r-1},$$

which is non-negative because the terms in it are in magnitude non-increasing. Hence (5 A) holds for a single element, and thus generally; (5 B) is proved similarly.

This form of the sieve is still comparatively easy to work with and has had a number of applications. From our point of view it will be particularly important because it will be used to develop the machinery of the enveloping sieve to which we shall advert briefly at a later stage in this chapter. In order, however, to illustrate our earlier remarks about the potential relevance of exclusion processes to situations that do not directly involve the sifting of sequences, we note that there has been an application of this sieve method to the treatment of intervals between consecutive members of sequences of integers, such intervals being characterized in part by the feature that no number within them is a member of the sequence (Hooley [34], [41]).

Nevertheless, even this form of the sieve is not sufficiently delicate to be an adequate tool for the investigation of the central sieving problems occurring in the theory of prime numbers. But in his second paper Brun [5] revealed the fundamental discovery that the power of the sieve method could be dramatically increased provided that it were directed towards the production of meaningful upper and lower bounds instead of asymptotic results. His method, which is largely combinatorial in nature and very complicated, has been later refined by other writers, and certain aspects of it still play a prominent part in current research on the field of study he initiated. The procedure involves the use of combinatorial identities already mentioned in conjunction with others, of which

$$N_{1,\,\ldots,\,n} = N - N(1) - \sum_{2 \leqslant r \leqslant n} N_{1,\,2,\,\ldots,\,r-1}(r) \qquad (8)$$

is an important example having a later application in this chapter. We do not need, however, to give any more details of this aspect of Brun's method, since apart from (8) our purposes will be better served by the simpler Selberg method that has superseded many of the earlier methods.

All the methods so far described depend on replacing the right-hand side of (1) by an appropriate sub-set of its terms. Yet Selberg [69], [70], [71] departed from this principle when he introduced his new and powerful upper bound method. Instead, his method, when interpreted in our general context, entailed the introduction of real numbers $\lambda_{j_1,\,\ldots,\,j_r}$ which were to be arbitrary for each sub-set $(j_1, \ldots, j_r)$ of $1, \ldots, n$ save that the value corresponding to the null sub-set was to be 1. For each element a in S there is then formed the expression

$$(\Sigma \lambda_{j_1,\,\ldots,\,j_r})^2,$$

where the sum is taken over all sub-sets $(j_1, \ldots, j_r)$ of the set of all indices corresponding to all the properties α_s that a possesses. This expression is never negative and is 1 if a is in the sifted (i.e. non-excluded) set. Thus the sum of this expression over all elements a in S gives an upper bound for $N_{1,\,\ldots,\,n}$. Next, transforming this sum and changing the order of summation, we obtain

$$N_{1,\,\ldots,\,n} \leqslant \sum_{\substack{(j_1,\,\ldots,\,j_r) \\ (k_1,\,\ldots,\,k_s)}} \lambda_{j_1,\,\ldots,\,j_r} \lambda_{k_1,\,\ldots,\,k_s} N(m_1, \ldots, m_t), \qquad (9)$$

where $(m_1, \ldots, m_t)$ is the set-theoretic union of $(j_1, \ldots, j_r)$ and $(k_1, \ldots, k_s)$ and where the summation is over all (ordered) pairs of sub-sets $(j_1, \ldots, j_r)$, $(k_1, \ldots, k_s)$ of $(1, \ldots, n)$. Then, provided we have a satisfactory knowledge of $N(m_1, \ldots, m_t)$, we can seek to obtain a good upper bound for $N_{1,\,\ldots,\,n}$ by finding the conditional minimum of the quadratic form on the right-hand side of (9). This method often gives surprisingly good results in practice.

We have taken the theory in its general context as far as is profitable for our purposes. Before, however, going on to explain specific applications of the theory, we should remark that all the formulae given above have obvious extensions in which each element of the set S is counted according to some positive weight. Thus, in particular, the set S may be replaced by a family of not necessarily distinct elements $a^{(1)}, \ldots, a^{(m)}$, as may be seen otherwise by introducing new properties A_s that are defined by the requirement that the superscript i have property A_s if and only if $a^{(i)}$ have property α_s.

2   Selberg's upper bound sieve

We take S now to be a set of positive integers and we take a set of distinct primes $p_1, \ldots, p_r$ to each of which there corresponds a property (such as, for example, divisibility by p_i). Let d denote, generally, a square-free number (possibly 1) composed entirely of prime factors taken from $p_1, \ldots, p_r$; then by uniqueness of factorization each sub-set of the above primes corresponds uniquely to a given number d, and conversely, where the empty set corresponds to 1 (the empty product). We can therefore use $N(d)$ to describe the number of integers in S having all the properties appertaining to the prime divisors of d, where $N = N(1)$. The Selberg lambda coefficients in like manner are appropriately redesignated by λ_d, where it is given that $\lambda_1 = 1$.

We now make the hypothesis, frequently realized in practice when N is large, that

$$N(d) = \frac{X}{f(d)} + R_d, \tag{10}$$

where $f(d)$ is multiplicative and $1 < f(p_i) < \infty$, and where R_d is small compared with $N(d)$ provided d is not too large (so X approximates to N). If $\lambda_d = 0$ for $d > z$, formula (9) then gives

$$X \sum_{d_1, d_2 \leqslant z} \frac{\lambda_{d_1} \lambda_{d_2}}{f([d_1, d_2])} + O\Big(\sum_{d_1, d_2 \leqslant z} |\lambda_{d_1}| \, |\lambda_{d_2}| \, |R_{[d_1, d_2]}| \Big) \tag{11}$$

as an upper bound for the number of sifted elements. Next, ignoring the remainder term for the time being, we seek a conditional minimum of the sum in the explicit term subject to the condition $\lambda_1 = 1$.† But this sum is

$$\sum_{d_1, d_2 \leqslant z} \frac{\lambda_{d_1} \lambda_{d_2}}{f(d_1) f(d_2)} f\{(d_1, d_2)\} = \sum_{d_1, d_2 \leqslant z} \frac{\lambda_{d_1} \lambda_{d_2}}{f(d_1) f(d_2)} \sum_{\substack{\rho|d_1 \\ \rho|d_2}} f_1(\rho), \tag{12}$$

† There is no longer an entirely obvious *a priori* reason for the existence of a conditional minimum, the justification in fact emerging *a posteriori* from (13).

in which
$$f_1(\rho) = \sum_{d|\rho} \mu(d) f(\rho/d) = f(\rho) \prod_{p|\rho} \left(1 - \frac{1}{f(p)}\right) > 0$$

by the Möbius inversion formula, all integers and subscripts here and below being restricted to be numbers not exceeding z that are of type d. Therefore (12) is

$$\sum_{\rho \leqslant z} f_1(\rho) u_\rho^2, \tag{13}$$

where
$$u_\rho = \sum_{\substack{d \leqslant z \\ d \equiv 0, \bmod \rho}} \frac{\lambda_d}{f(d)},$$

and where the equations adjoint to these are

$$\frac{\lambda_d}{f(d)} = \sum_\sigma \mu(\sigma) u_{d\sigma} \tag{14}$$

by one of the Möbius formulae. The condition $\lambda_1 = 1$ being equivalent to

$$\sum_{\rho \leqslant z} \mu(\rho) u_\rho = 1,$$

the conditional minimum is easily seen to be

$$\frac{1}{V(z)} \tag{15}$$

by (13), where
$$V(z) = \sum_{\rho \leqslant z} \frac{\mu^2(\rho)}{f_1(\rho)};$$

furthermore this minimum is attained when

$$u_\rho = \frac{\mu(\rho)}{V(z) f_1(\rho)}. \tag{16}$$

Since the values of λ_d corresponding to (16) are then easily inferred from (14), we can then estimate an upper bound for the remainder term in (11). Finally we must choose z so as to make the explicit term as small as is possible without the remainder term becoming preponderant. Thus with an appropriate value of z we may expect to obtain

$$\frac{X(1+\eta)}{V(z)} \tag{17}$$

as an upper bound for the number of sifted elements.

It is appropriate to make several comments on this procedure before it is applied to special cases. First the process is only influenced by that part of the sifting that appertains to the relevant primes not exceeding z. Next, for any sifting prime p, the inequality $N(p) \leqslant N$ together with the given formula for $N(p)$ imply the assumed inequality for $f(p)$ in the weaker form $f(p) \geqslant 1$. Here the import of the excluded cases $f(p) = 1$ and $f(p) = \infty$ is such that they need not feature directly in the structure of the method. The former case corresponds to the situation in which all or nearly all members of the set have the property associated with p, it then being obvious that the sifting with respect to p results in an empty or nearly empty set. At the other extreme, the sifting in terms of p becomes superfluous when $f(p) = \infty$, because in that case it is already given that hardly any members of the set have the property in question.

It is also helpful to elaborate on two matters of a quantitative nature. Assuming that the procedure is still limited to those primes p for which $f(p) \neq 1, \infty$, one is always led to the conclusion that

$$|\lambda_d| \leqslant 1 \tag{18}$$

by means of the identity

$$V(z)\,\lambda_d = \mu(d) \prod_{p \mid d} \left(1 - \frac{1}{f(p)}\right)^{-1} \sum_{\substack{\rho' \leqslant z/d \\ (\rho',\, d)=1}} \frac{\mu^2(\rho')}{f_1(\rho')}$$

$$= \mu(d) \sum_{\delta \mid d} \frac{\mu^2(\delta)}{f_1(\delta)} \sum_{\substack{\rho' \leqslant z/d \\ (\rho',\, d)=1}} \frac{\mu^2(\rho')}{f_1(\rho')}$$

that is implied by (14) and (16). In some cases a much sharper inequality holds; for example, $f(p) = 2$ would give

$$\lambda_d = O(d^{-1+\epsilon}).$$

Finally, by writing

$$\frac{1}{f_1(\rho)} = \frac{1}{f(\rho)} \prod_{p \mid \rho} \left(1 - \frac{1}{f(\rho)}\right)^{-1}$$

and expanding the product through the binomial theorem, one can infer the useful inequality

$$V(z) \geqslant \sum_{\delta \leqslant z} \frac{1}{f(\delta)}, \tag{19}$$

where δ denotes generally any number (square-free or otherwise) whose prime factors belong to the sifting set of primes and where the definition of $f(n)$ is extended to non square-free numbers by taking it to be totally multiplicative.

We give two applications that will be needed later. The first is an upper bound for $\pi(x; a, k)$ where $(a, k) = 1$ and $k \leqslant x$. We consider those numbers $n \equiv a$, mod k, not exceeding x that are not divisible by primes p such that $p \leqslant z$ and $p \nmid k$ (cf. remark in earlier paragraph). Here the formula (10) takes the form

$$N(d) = \frac{x}{kd} + O(1),$$

and the upper bound becomes

$$\frac{x}{kV(z)} + O(z^2)$$

on minimizing the first term in (11). In this, by (19),

$$\frac{k}{\phi(k)} V(z) \geqslant \prod_{p|k} \left(1 - \frac{1}{p}\right)^{-1} \sum_{\substack{(k,\,m)=1 \\ m \leqslant z}} \frac{1}{m} \geqslant \sum_{\mu \leqslant z} \frac{1}{\mu} \sim \log z.$$

Hence, choosing $z = (2x/k)^{\frac{1}{2}-\epsilon}$, we obtain an upper bound to which must be added $\pi(z; a, k) = O(z)$ in order to furnish an upper bound for $\pi(x; a, k)$ itself. This yields a form of the Brun–Titchmarsh theorem, which we state as Lemma 1.

LEMMA 1   (*The Brun–Titchmarsh Theorem*)   *For* $(a, k) = 1$ *and* $k \leqslant x$, *we have*

$$\pi(x; a, k) < \frac{(2+\eta)\,x}{\phi(k)\log\,(2x/k)} \quad (x > x_0(\eta)).$$

Although this formulation of the lemma suffices for the applications in this tract, it is of interest to note in passing that a different method due to Montgomery and Vaughan [58] has shewn the presence of the arbitrarily small number η in the result to be unnecessary. Alternatively, as remarked by Selberg, the number η may be removed by attending painstakingly to a number of details that are implicit in the treatment outlined above.

The lemma should be compared with Bombieri's theorem, to which we shall also allude from time to time. The latter theorem, a special form of which is stated in Chapter 5, states, roughly speaking, that the result of Lemma 1 can be improved to $\pi(x;\ a, k) \sim \mathrm{li}\, x/\phi(k)$ for almost all k not exceeding $x^{\frac{1}{2}} \log^{-A} x$.

As the second application we prove

LEMMA 2 *For any $r < \frac{1}{2}n$, the number of solutions of*

$$n - p = p'r; \; p < n$$

in (positive) primes p, p' is

$$O\left(\frac{n^2}{\phi(nr)\log^2(n/r)}\right).$$

Since the cases $(n, r) > 1$ and nr odd are each trivial, we can assume in what follows that $(n, r) = 1$ and that nr is even. We consider the number of solutions in l, m of

$$n - l = mr; \; l < n$$

for which neither l nor m is divisible by primes $p \leqslant z$. But this condition on l, m is equivalent to the stipulation that m shall not satisfy either

$$m \equiv 0, \bmod p, \quad \text{or} \quad mr \equiv n, \bmod p, \tag{20}$$

i.e. if $p \nmid nr$, then m shall not belong to two residue classes, mod p; whereas, if $p \mid nr$, then m shall not belong to one residue class, mod p, the second congruence in (20) either being impossible or becoming equivalent to the first. Therefore, defining $f(d)$ by multiplicativity and by

$$f(p) = \begin{cases} \dfrac{p}{2}, & \text{if} \quad p \nmid nr, \\[2mm] p, & \text{if} \quad p \mid nr, \end{cases}$$

so that the condition $f(d) > 1$ obtains, we deduce that

$$N(d) = \frac{n}{rf(d)} + O(d^\epsilon),$$

where $N(d)$ is the number of integers m less than n/r that satisfy one or other of the congruences in (20) for each prime divisor p of d. The remainder of the argument being similar to its counterpart in Lemma 1, it is enough here to observe that

$$\prod_{p \mid nr}\left(1 + \frac{1}{p}\right) \cdot V(z) \geqslant \sum_{m \leqslant z} \frac{\mu^2(m)\, 2^{\omega(m)}}{m} > A_1 \log^2 z$$

and hence that

$$V(z) > \frac{A_2 \phi(nr) \log^2 z}{nr},$$

in which an appropriate choice of z is $(n/r)^{\frac{1}{2}-\varepsilon}$.

3 The lower bound sieve

No effective lower bound sieve method so far discovered approaches the Selberg upper bound method in simplicity. Usually the method depends on linking the Selberg upper bound sieve with Brun's lower bound sieve, the details becoming very intricate if sharp results are needed. We are content here to describe a simple version of the method by means of an illustrative example that we shall be able to utilize in Chapter 6.

We use the method to elicit a positive lower bound for the number, $M(x,v)$, of positive integers not exceeding x that are not divisible by primes p such that $p \leqslant v$ and $p \equiv 1$, mod 4. Let q be any prime number that is congruent to 1, mod 4. Then, defining $N(x,q)$ to be the number of positive integers not exceeding x that are divisible by q but not divisible by any prime p such that $p < q$ and $p \equiv 1$, mod 4, we have by (8) that

$$M(x,v) = [x] - \sum_{\substack{q \leqslant v \\ q \equiv 1, \text{ mod } 4}} N(x,q). \tag{21}$$

If $N(x,q)$ is replaced in the right-hand side of this by an upper bound given by a suitable use of Selberg's sieve method, we shall obtain a lower bound for $M(x,v)$. Taking $z = v^\alpha$ for a suitable constant $\alpha > 1$ in the application of the Selberg method to $N(x,q)$, we obtain, by (11), (18), and (19),

$$N(x,q) < \frac{x}{q \sum_{n_q \leqslant v^a} \frac{1}{n_q}} + O(v^{2\alpha}), \tag{22}$$

where n_q denotes a number all of whose prime factors are less than q and are congruent to 1, mod 4. Also, we have

$$\sum_{n_q \leqslant v^a} \frac{1}{n_q} = \sum_{n_q} \frac{1}{n_q} - \sum_{n_q > v^a} \frac{1}{n_q} = \prod_{\substack{p < q \\ p \equiv 1, \text{ mod } 4}} \left(1 - \frac{1}{p}\right)^{-1} - \sum_{n_q > v^a} \frac{1}{n_q}. \tag{23}$$

We require an inequality for the second term in the right-hand side of (23), one that meets our needs here being easily obtained thus. Let $y > q$. Then, for $\sigma > 0$,

$$\sum_{n_q > y} \frac{1}{n_q} \leqslant y^{-\sigma} \sum_{n_q} \frac{1}{n_q^{1-\sigma}} = y^{-\sigma} \prod_{\substack{p < q \\ p \equiv 1, \bmod 4}} \left(1 - \frac{1}{p^{1-\sigma}}\right)^{-1}$$

from which, by setting $\sigma = \dfrac{1}{\log q} < 1$, we can infer that

$$\sum_{n_q > y} \frac{1}{n_q} < A_3 \sqrt{(\log q)} \exp\left(-\frac{\log y}{\log q}\right) \tag{24}$$

since now

$$\log \prod_{\substack{p < q \\ p \equiv 1, \bmod 4}} \left(1 - \frac{1}{p^{1-\sigma}}\right)^{-1} = \sum_{\substack{p < q \\ p \equiv 1, \bmod 4}} \frac{1}{p^{1-\sigma}} + O\left(\sum_{p<q} \frac{1}{p^{2(1-\sigma)}}\right)$$

$$= \sum_{\substack{p < q \\ p \equiv 1, \bmod 4}} \frac{1}{p} \exp\left(\frac{\log p}{\log q}\right) + O(1)$$

$$= \sum_{\substack{p < q \\ p \equiv 1, \bmod 4}} \frac{1}{p} + O\left(\frac{1}{\log q} \sum_{p<q} \frac{\log p}{p}\right) + O(1)$$

$$= \tfrac{1}{2} \log \log q + O(1).$$

We obtain from (22), (23), and (24) that

$$N(x, q) < \frac{x}{q} \prod_{\substack{p < q \\ p \equiv 1, \bmod 4}} \left(1 - \frac{1}{p}\right) + \frac{A_4 x}{q \sqrt{(\log q)}} \exp\left(-\frac{\log v^\alpha}{\log q}\right) + O(v^{2\alpha})$$

for $\alpha > \alpha_0(A_3)$ and $q \leqslant v$, and then from this and (21) that

$$M(x, v) > x \left(1 - \sum_{\substack{q \leqslant v \\ q \equiv 1, \bmod 4}} \frac{1}{q} \prod_{\substack{p < q \\ p \equiv 1, \bmod 4}} \left(1 - \frac{1}{p}\right)\right)$$

$$- \frac{A_5 x}{\alpha^{\frac{3}{2}} \log^{\frac{3}{2}} v} \sum_{q \leqslant v} \frac{\log q}{q} + O(v^{2\alpha+1})$$

$$> x \prod_{\substack{p \leqslant v \\ p \equiv 1, \bmod 4}} \left(1 - \frac{1}{p}\right) - \frac{A_6 x}{\alpha^{\frac{3}{2}} \log^{\frac{1}{2}} v} + O(v^{2\alpha+1})$$

$$> \frac{A_7 x}{\log^{\frac{1}{2}} v} - \frac{A_6 x}{\alpha^{\frac{3}{2}} \log^{\frac{1}{2}} v} + O(v^{2\alpha+1}). \tag{25}$$

In this now choose α so that also $\alpha > (A_6/A_7)^{\frac{2}{3}}$ and then choose v to be x^β, where $\beta < 1/(2\alpha + 1)$. We thus deduce that

$$M(x, v) > \frac{Ax}{\log^{\frac{1}{2}} v} \tag{26}$$

for $v \leqslant x^\beta$ and $x > x_0$.

Arguments of this sort are capable of considerable refinement; they are powerful enough, for example, to provide an elementary (in the technical sense) proof of the theorem that all sufficiently large even numbers are the sum of two numbers, one having not more than two prime factors and the other not more than three prime factors. Moreover, by using Bombieri's theorem in conjunction with another elaborate development of these methods, Chen Jing-run has recently proved that all sufficiently large even numbers are the sum of a prime and a number having at most two prime factors, the reader being referred to the monograph of Halberstam and Richert [25] for further details of this important contribution to the study of Goldbach's problem.

As a matter of fact, in its application to the particular problem just considered, our method turns out to be greatly inferior to transcendental methods, which are capable of estimating $M(x, x)$ (the case of greatest interest) asymptotically; nevertheless, our method has established a fact that will be of the utmost importance in Chapter 6.

To elicit this fact we must reinterpret some of the previous proof. An examination of (21) together with Selberg's method reveals that we are finding a lower bound for $M(x, v)$ by counting each integer n not exceeding x with a weight equal to

$$1 - \sum_{\substack{q \leqslant v \\ q \equiv 1, \, \text{mod } 4}} \big(\sum_{qd_q | n} \lambda_{qd_q} \big)^2, \tag{27}$$

where d_q denotes a square-free number not exceeding v^α composed entirely of prime factors p such that $p < q$ and $p \equiv 1, \text{mod } 4$, and where $\lambda_q = 1$. Furthermore, either from general principles or by a direct examination of (27), it is readily seen that this weight is 1 for any number n whatsoever that is not divisible by the sifting set of primes but is non-positive for any number n whatsoever that is divisible by at least one prime in the sifting set.

Letting δ denote, generally, any square-free number composed of primes p such that $p \leqslant v$ and $p \equiv 1$, mod 4, we summarize this fact by a statement of the form

$$\sum_{\delta|n} \rho(\delta) \leqslant \sum_{\delta|n} \mu(\delta), \tag{28}$$

where the left-hand side of (28) is expressed as a sum over divisors of n and where $\rho(\delta) = \rho_{v,\alpha}(\delta)$. In this $\rho(\delta) = 0$ for $\delta > v^{2\alpha+1}$, and also $\rho(\delta) = O(\delta^\epsilon)$ after the quadratic forms have been minimized. Also, since in this notation

$$M(x,v) = \sum_\delta \rho(\delta) \left[\frac{x}{\delta} \right],$$

we have here that

$$\sum_\delta \frac{\rho(\delta)}{\delta} > \frac{A}{\log^{\frac12} v} \tag{29}$$

by (26).

We end this section by making a few informal remarks about sieving limits, the theory of which has been considered in detail by Selberg [72]. For a full description of this theory, particular cases of which we shall have occasion to mention in the sequel, the reader is referred to Selberg's paper on the matter. Here we merely attempt to summarize the position by saying that the power of the sieve method is constrained by definite limits provided certain natural assumptions are made concerning the circumstances of the application. Characteristically, in fact, the use of the sieve method is circumscribed by the following factors: (i) there is a formula of type (10) in which some bound for R_d is given, (ii) the effect of R_d in the estimations can only be assessed through the use of its absolute value, (iii) the sieving is performed through the use of a divisor sum such as appears in (28).

Consider, as a specific case, the analogue of the problem just considered in which multiples of *all* primes not exceeding v are excluded from the numbers not exceeding x, where $R_d = O(1)$ is the only property of R_d that is to be utilized. A similar method will yield a lower bound of the form $Ax/\log v$ provided v does not exceed a sufficiently small power of x. Here Selberg's theory shews that a refinement in the method will give a comparable result for $v \leqslant x^{\frac12-\epsilon}$ but that it cannot give it for $v > x^{\frac12+\epsilon}$. Yet, if the sifting set of primes is thinner, then we may expect, not unnaturally, to obtain positive lower bounds for a wider range of v. For example, it can be shewn that (26) can in fact be obtained for $v \leqslant x^{1-\epsilon}$ by an improved version of our method; here, loosely speaking, we are only sifting with respect to one half of the set of primes, whence the term '$\frac{1}{2}$-residue sieve' that is applied to the method in this and similar contexts. The '$\frac{1}{2}$-residue sieve' is naturally also involved if we sift instead with respect to primes congruent to 3, mod 4.

Actually, in the particular context just described, it even transpires that it is often feasible to derive an asymptotic formula for the case $v = x$ itself, provided that the set of sieving primes is appreciably less numerous than that used in the $\frac{1}{2}$-residue sieve. Already obvious from our earlier remarks on the simple asymptotic sieve when the sifting set is very sparse, this assertion is more generally substantiated by the method of § 1, paragraph 5, since the number of terms in (1) is then limited to

$$O(x \log^{-\frac{1}{2}-\delta} x)$$

by the nature of the application. Yet these asymptotic formulae are only possible because of special features that are inherent in this particular case, the method ceasing to be viable if we consider similar problems about the sifting of integers in an interval $(H, H + x)$, where H is large compared with x. In contrast, the above results on sieving limits are not affected by such a change in the problem, all assumptions previously made in respect of (i), (ii), and (iii) being still applicable.

A major change in the circumstances of application naturally affects the best possible results obtainable by the theory of sieving limits. For example, if we attempt to sieve the sequence $p - 1$ for $p \leqslant x$, then Bombieri's theorem would imply that we could assume a formula of type (10) in which R_d on average was about $O(x^{\frac{1}{2}})$. Here, therefore, the sieving limits would be less, and it is not hard to see on the basis of what has already been stated that, in the application of the $\frac{1}{2}$-residue sieve to this sequence, the sieving limit† would be $v = x^{\frac{1}{2}-\epsilon}$.

---

† Our definition of sieving limit differs slightly from that used by Selberg.

4 The large sieve

The large sieve arises in connection with the problem of estimating the number of integers not exceeding x which do not belong to $w(p)$ assigned residue classes, mod p, for any $p \leqslant z$. An upper bound here can be derived by using Selberg's method in much the same way as it was applied to the examples in § 2. Now formula (10) takes the form

$$N(d) = \frac{xw(d)}{d} + O\{w(d)\}, \tag{30}$$

where $f(d) = \dfrac{d}{w(d)}$ and $w(d)$ is defined for appropriate square-free numbers d through multiplicativity. Then, by following the analysis of § 2, we obtain an upper bound of the form

$$\frac{x}{V(z)} + R(x, z), \tag{31}$$

where, defining $v(d)$ by $v(p) = p - w(p)$ and by multiplicativity, we have

$$V(z) = \sum_{d \leqslant z} \frac{\mu^2(d)\, w(d)}{v(d)},$$

and where $R(x, z)$ is a remainder term. When $w(p)$ is effectively bounded it is easy to shew that the first term in (11) dominates for $z < x^{\frac{1}{2}-\epsilon}$. When, however, $w(p)$ is large the method is less satisfactory because smaller values of z need to be chosen in order that $R(x, z)$ shall not preponderate. For instance, if $w(p) = \frac{1}{2}(p-1)$ for $p > 2$, then it is necessary to take $z = x^{\frac{1}{3}-\epsilon}$, the corresponding bound being then

$$O(x^{\frac{2}{3}+\epsilon}). \tag{32}$$

An alternative approach to the problem is available through the use of inequalities of the large sieve type. By their means Montgomery [57] has proved an upper bound of type (31) in the improved form

$$\frac{x + Az^2}{V(z)}, \tag{33}$$

his treatment having been subsequently presented in various

alternative forms by other writers (see, in particular, [58] and the references given therein). This method not only gives better results than (31) when $w(p)$ is large but it also obviates the complicated calculations inherent in the estimation of $R(x, z)$. It can be applied to any sequence $f(n)$ in which the sifting corresponds to the sifting of residue classes from the argument sequence n. For example, the Brun–Titchmarsh theorem can be proved equally well by the Montgomery method on observing that non-divisibility of $a + rk$ by any prime p not dividing k is equivalent to $r \not\equiv -a\bar{k}$, mod p, where $k\bar{k} \equiv 1$, mod p.

Notwithstanding the advantages of this alternative approach, the great merit of the Selberg method is in its applicability to a very wide range of sequences. Also, contrary to what has been implied by some writers, the Selberg method can effectively handle the exclusion of many residue classes for each relevant prime, even though the alternative approach may initially yield better estimates in the situations to which it is applicable. Furthermore, the author recently has ascertained that the original Selberg approach itself will almost yield (33) provided the method is suitably refined. We discuss briefly the way in which the latter improvement is obtained, since this will serve as an introduction to some later remarks we wish to make.

The underlying idea is that we may expect on average the remainder term in (30) to be $O\{w^{\frac{1}{2}}(d)v^{\frac{1}{2}}(d)/d^{\frac{1}{2}}\}$. Also, if we could simply use the latter estimate for each d, then (33) could be substantiated without difficulty. Although a direct substitution of this type would not in the general case be legitimate, the consequence of such a substitution can be shewn to be true by considering the remainder term in mean square. For this purpose we prepare the remainder term by letting $h(d)$ denote the $w(d)$ residue classes, mod d, that give rise to (30). Then, since the contribution of each single residue class to $N(d) - \dfrac{xw(d)}{d}$ can be expressed in terms of the Fourier series for $[u] - u + \frac{1}{2}$ by a method that is similar to one used in Chapter 2, it can be concluded that the remainder term in (31) can be made to depend on sums of the form

$$\sum_{\nu \in h(d)} e^{2\pi i l \nu / d}. \tag{34}$$

As for these sums, they are comparatively easily handled in mean square because they have a multiplicative property.

Complicated though this method may be, it provides a basis for applying the sieve method in situations where the other methods so far discussed appear to be entirely unsuitable. A simple prototype of the method is provided by the special case of the above example in which the quadratic residues are eliminated for each prime p, expression (34) becoming in this case a Gauss sum with a known non-trivial estimate.† In more complicated cases the use of the sieve method proves to be possible only if the error terms involved can be estimated through the use of similar but more recondite exponential sums that may involve several variables of summation. We mention two such cases in Chapter 4.

We use the term large sieve to mean a sieve that involves in some way or other the elimination of a large number of residue classes, mod p, for a set of primes p. We shall not need here to develop the Montgomery method or to expand further our remarks on adapting the Selberg method for the large sieve case. Instead it will be enough to use the following theorem of Gallagher [21], which is proved in a short and ingenious manner. Indeed for many problems Gallagher's theorem serves as well as the results that require so much more preparation for their proof.

LEMMA 3 *If a set of positive integers not exceeding x includes only representatives from at most $v(k) > 0$ residue classes for each prime-power modulus k, then the number of integers in the set is at most*

$$\Big(\sum_{k \in T} \Lambda(k) - \log x \Big) \Big/ \Big(\sum_{k \in T} \frac{\Lambda(k)}{v(k)} - \log x \Big)$$

whenever T is any finite set of moduli for which the denominator is positive.

For the proof let Z be the number of integers in the set, let $Z(h, k)$ be the number of these that are congruent to h, mod k, and let a denote a typical integer in the set. Then

$$Z^2 = \Big(\sum_{h=1}^{k} Z(h, k) \Big)^2 \leqslant v(k) \sum_{h=1}^{k} Z^2(h, k)$$

† Here it can be shewn by other methods that the sifted set is actually empty if z is sufficiently large in terms of x.

by the Cauchy–Schwarz inequality. Hence

$$\frac{Z^2}{v(k)} \leqslant \sum_{a' \equiv a'', \bmod k} 1 = Z + \sum_{\substack{k \mid (a'-a'') \\ a' \neq a''}} 1,$$

and so

$$Z^2 \sum_{k \in T} \frac{\Lambda(k)}{v(k)} \leqslant Z \sum_{k \in T} \Lambda(k) + \sum_{a' \neq a''} \sum_{k \mid (a'-a'')} \Lambda(k)$$

$$\leqslant Z \sum_{k \in T} \Lambda(k) + \log x \, (Z^2 - Z),$$

from which the result follows immediately.

5   The enveloping sieve

Since the concept of the enveloping sieve is discussed in the introduction to Chapter 5, we confine ourselves to a few initial comments. Sieve methods are usually applied to problems in such a way that each sequence involved is investigated through the use of an individual sieving function that is defined in terms of the particular characteristics of that sequence (cf., in particular, the discussion on Selberg's sieve method). However, there are problems involving classes of sequences in which it is advantageous or, indeed, necessary to use the same exclusion process for each sequence in the class. It is to the sieve method when used in the latter manner that we give the term enveloping sieve.

One application of the idea is to the Brun–Titchmarsh theorem. In the proof of this theorem we used a sieve that depended on the particular arithmetic progression in question (actually the dependence is only on the modulus k and on x). However, by using an invariant sieve, it is possible to shew that in various useful senses the Brun–Titchmarsh theorem can be improved on average (Hooley [40], [44]), it being no essential part of the method that the lambda coefficients must correspond to a conditional minimum of a quadratic form. We remark also that these developments are achieved through a technique that enables us to consider non-trivially the influence of the remainder term R_d in (10) on the estimations (see also Motohashi [60] and Goldfeld [22]). Here again, we have an example where the flexibility of the Selberg method shews to advantage.

6 The normal order method

Reverting to the general context as described in the first section, we conclude by mentioning that there are situations in which our methods fail because of our inability to calculate $N(j_1, \ldots, j_s)$ for the larger values of s. The following method, however, is sometimes available provided we can estimate $N(j)$ and $N(i, j)$ satisfactorily.

Let $r \leqslant n$ and define $\nu(a)$ for each $a \in S$ to be the total number of indices not exceeding r that correspond to properties enjoyed by a. Then we attempt to compare

$$\sum_{a \in S} \nu(a) = \sum_{1 \leqslant j \leqslant r} N(j)$$

with

$$\sum_{a \in S} \nu^2(a) = \sum_{1 \leqslant j \leqslant r} N(j) + 2 \sum_{1 \leqslant i < j \leqslant r} N(i, j).$$

If in some asymptotic sense appropriate to the problem and the context we can shew that

$$\frac{1}{N} \sum_{a \in S} \nu(a) \to \infty$$

and

$$\left(\frac{1}{N} \sum_{a \in S} \nu(a) \right)^2 \sim \frac{1}{N} \sum_{a \in S} \nu^2(a),$$

we can conclude that $\nu(a)$ has a *normal order* that is large. In particular we would then deduce that

$$N_{1, \ldots, n} = o(N),$$

since elements a without any of the properties would certainly then be exceptional.

An example of this method is given by Turán's work ([29], Chapter 22) on the normal order of $\omega(n)$, which implicitly supplies an entirely elementary and easy proof of the inequality

$$\pi(x) = O\left(\frac{x}{\log \log x} \right).$$

Another application of this idea is given in Chapter 7.

A somewhat similar procedure is available through a rudimentary form of Selberg's method that has been analyzed by the author. Here the structure of the method described at the end of §1 is retained save that the extra condition $\lambda_{j_1, \ldots, j_r} = 0$ is imposed for $r > 1$, the quadratic form in (9) now having coefficients that are of the form $N(j)$ and $N(i, j)$ only. In the context of §2 the ensuing conditional minimization becomes particularly easy to implement and leads to the formal main term $\dfrac{X}{W(z)}$ corresponding to the values

$$\lambda_p = -\frac{f(p)}{f_1(p)\, W(z)},$$

where

$$W(z) = 1 + \sum_{p_i \leqslant z} \frac{1}{f_1(p_i)}.$$

This method, which also has applications to situations of large sieve type, may be regarded as standing in the same relationship to Selberg's method as do some of the precursors of Montgomery's result (33) to the latter result itself (*vide*, references made in [58] to the work of Linnik, Rényi, and Gallagher).

2. On a problem of Chebyshev – an application of the Selberg sieve method

1 Introduction

The interest of several mathematicians has been attracted by Chebyshev's theorem to the effect that, if P_x is the greatest prime factor of

$$\prod_{n \leqslant x} (n^2 + 1),$$

then

$$\frac{P_x}{x} \to \infty$$

as $x \to \infty$. Revealed posthumously as little more than a fragment in one of Chebyshev's manuscripts, the theorem was first published and fully proved in a paper by Markov [54] in 1895, while later in the same year a generalization by Ivanov [46] appeared in which the polynomial $n^2 + 1$ was replaced by $n^2 - D$ for any negative D (an account of the work of both these writers is to be found in paragraphs 147 and 149 of Landau's *Primzahlen* [49]). In 1921 Nagell [61] improved and further generalized Chebyshev's theorem by shewing that for any $\epsilon < 1$ and for all sufficiently large x

$$\frac{P_x}{x} > \log^\epsilon x,$$

where P_x is the largest prime factor in the product obtained by replacing $n^2 + 1$ by any irreducible integral polynomial $f(n)$ of degree greater than 1. Another development was due to Erdös [14], who in 1952 improved Nagell's result by shewing that

$$\frac{P_x}{x} > (\log x)^{A_1 \log\log\log x}$$

by a method which he stated could be refined further to yield

$$\frac{P_x}{x} > e^{\log^{A_2} x}$$

for any $A_2 < 1$.

However, by combining Chebyshev's method with a somewhat unusual application of the upper bound sieve method, the author [37] proved in 1967 that

$$P_x > x^{\frac{11}{10}}$$

for any irreducible quadratic polynomial $n^2 - D$ with D positive or negative. The successful use of the sieve method here depended on a theory of exponential sums that was derived from the theory of the representation of numbers by binary quadratic forms of (Gaussian) determinant D. Because it has not yet been possible to generalize this theory of exponential sums satisfactorily, the method in its present form can only be extended to cover all irreducible quadratic forms.

We shall describe this application of the sieve method by proving the above result for the special polynomial $n^2 + 1$ that was considered by Chebyshev. The restriction of the value of D to -1 has been made to simplify the details, the principles behind the method being in no way affected.

At the end we shall make some comments on the relevance of the method to other problems, including some that will be taken up again at a later stage in this tract.

2   Development of the method

In the very beginning the method follows that of Chebyshev and Markov. If $N_x(l)$ is the number of positive integers n not exceeding x with the property that $n^2 + 1$ be divisible by l, we have

$$\prod_{\substack{p \leqslant P_x \\ p^a \leqslant x^2+1}} p^{N_x(p^a)} = \prod_{n \leqslant x} (n^2 + 1) > ([x\,!])^2,$$

where the left-hand side is taken over primes p and positive values of α. Consequently, by Stirling's theorem,

$$\sum_{\substack{p \leqslant P_x \\ p^a \leqslant x^2+1}} N_x(p^\alpha) \log p > 2x \log x + O(x). \tag{35}$$

Next $\displaystyle\sum_{\substack{p \leqslant P_x \\ p^a \leqslant x^2+1}} N_x(p^\alpha) \log p = \sum_{x < p \leqslant P_x} N_x(p) \log p + \sum_{p \leqslant x} N_x(p) \log p$

$$+ \sum_{\substack{p^a \leqslant x^2+1 \\ \alpha > 1}} N_x(p^\alpha) \log p$$

$$= \Sigma_A + \Sigma_B + \Sigma_C, \quad \text{say,} \tag{36}$$

the condition $p \leqslant P_x$ being omitted from Σ_C as it is in reality superfluous. The lower bound for Σ_A required for the application of our method can be inferred from (35) and (36) through upper bounds we now derive for Σ_B and Σ_C.

The estimates for Σ_B and Σ_C are formed by considering an expression for $N_x(l)$. We have

$$N_x(l) = \sum_{\substack{n^2+1\equiv 0,\, \mathrm{mod}\, l \\ n \leqslant x}} 1 = \sum_{\substack{\nu^2+1\equiv 0,\, \mathrm{mod}\, l \\ 0<\nu\leqslant l}} \sum_{\substack{n\equiv\nu,\, \mathrm{mod}\, l \\ n\leqslant x}} 1$$

$$= \sum_{\substack{\nu^2\equiv -1,\, \mathrm{mod}\, l \\ 0<\nu\leqslant l}} \left(\left[\frac{x-\nu}{l} \right] - \left[\frac{-\nu}{l} \right] \right), \quad (37)$$

which formula will be used later to develop another expression for $N_x(l)$. Here it suffices to deduce that

$$N_x(l) = \frac{x\rho(l)}{l} + O\{\rho(l)\}, \quad (38)$$

where $\rho(l)$ is the number of roots of the congruence $\nu^2 \equiv -1, \mathrm{mod}\, l$. Now $\rho(p)$ is 2 if $p \equiv 1, \mathrm{mod}\, 4$, is 0 if $p \equiv 3, \mathrm{mod}\, 4$, and is 1 if $p = 2$; also $\rho(p^\alpha) = O(1)$ always. Therefore, by (36) and (38),

$$\Sigma_B = x \sum_{p\leqslant x} \frac{\rho(p)\log p}{p} + O(\sum_{p\leqslant x} \rho(p)\log p) = 2x \sum_{\substack{p\leqslant x \\ p\equiv 1,\, \mathrm{mod}\, 4}} \frac{\log p}{p} + O(x)$$

$$+ O(\sum_{p\leqslant x} \log p)$$

$$= x\log x + O(x). \quad (39)$$

Also

$$\Sigma_C = O\left(\sum_{p\leqslant (x^2+1)^{\frac{1}{2}}} \log p \sum_{2\leqslant\alpha\leqslant\{\log (x^2+1)\}/\log p} \left\{ \frac{x}{p^\alpha} + 1 \right\} \right)$$

$$= O\left(x \sum_p \frac{\log p}{p(p-1)} \right) + O(\log x \sum_{p\leqslant 2x} 1) = O(x). \quad (40)$$

The lower bound

$$\Sigma_A > x\log x + O(x) \quad (41)$$

is obtained immediately from (35), (36), (39), and (40). On the other hand Σ_A is a particular example of sums of the form

$$T_x(y) = \sum_{x<p\leqslant y} N_x(p)\log p,$$

for which we shall shew that an upper bound can be derived in a complicated manner by Selberg's sieve method. The comparison of this upper bound with the lower bound given by (41) will yield the required lower estimate for P_x. It is in this determination of the upper bound and in the subsequent comparison that the more individual aspects of the method are presented.

We prepare $T_x(y)$ for the application of the sieve method, assuming throughout that y in this sum (but not necessarily elsewhere) is subject to the restriction $x^{\frac{12}{11}} < y < x^2$. A single application of the sieve method to estimate the complete sum directly does not give the optimum result, since the sum can be divided into a series of segments for which we can derive individual upper estimates which, while being of a common order of magnitude, are nevertheless of variable precision. Each such segment can be estimated by two different methods, the position of the segment in the sum determining which method is the more favourable. First then, dividing the sum into two parts that correspond to the methods to be used, we write

$$T_x(y) = \sum_{x < p \leqslant xX} N_x(p) \log p + \sum_{xX < p \leqslant y} N_x(p) \log p$$
$$= T_x(xX) + T'_x(y), \quad \text{say}, \quad (42)$$

where $X = x^{\frac{1}{11}}$. The sums $T_x(xX)$ and $T'_x(y)$ are then each to be split up into segments as follows.

To consider the first sum let

$$V_x(v) = \sum_{v < p \leqslant ev} N_x(p).$$

Then

$$T_x(xX) \leqslant \sum_{0 \leqslant \alpha < \log X} \sum_{xe^\alpha < p \leqslant xe^{\alpha+1}} N_x(p) \log p$$
$$\leqslant \sum_{0 \leqslant \alpha < \log X} \log (xe^{\alpha+1}) V_x(xe^\alpha). \quad (43)$$

To effect the corresponding division of $T'_x(y)$ it is necessary to transform the sum. By the definition of $N_x(p)$ we have

$$T'_x(y) = \sum_{\substack{xX < p \leqslant y \\ pm = n^2+1 \\ n \leqslant x}} \log p = \sum_{m > x^2/y \log^8 x} + \sum_{m \leqslant x^2/y \log^8 x}$$
$$= T''_x(y) + T'''_x(y), \quad \text{say}, \quad (44)$$

the conditions of summation for $T_x'''(y)$ implying that

$$n < (pm)^{\frac{1}{2}} \leqslant x \log^{-4} x$$

and that $m \leqslant x \log^{-4} x$. Therefore

$$T_x'''(y) \leqslant 2 \log x \sum_{\substack{lm=n^2+1 \\ m,\, n \leqslant x \log^{-4} x}} 1 = 2 \log x \sum_{m \leqslant x \log^{-4} x} N_{x \log^{-4} x}(m),$$

whence, by (38),

$$T_x'''(y) \leqslant \frac{2x}{\log^3 x} \sum_{m \leqslant x \log^{-4} x} \frac{\rho(m)}{m} + O\Big(\log x \sum_{m \leqslant x \log^{-4} x} \rho(m)\Big)$$

$$= O\Big(\frac{x}{\log^3 x} \sum_{m \leqslant x \log^{-4} x} \frac{\rho(m)}{m}\Big) = O\Big(\frac{x}{\log^3 x} \sum_{m \leqslant x} \frac{d(m)}{m}\Big)$$

$$= O\Big(\frac{x}{\log x}\Big). \tag{45}$$

Next, since the conditions of summation in $T_x''(y)$ imply that $m \leqslant ex/X$, we have

$$T_x''(y) \leqslant \sum_{\substack{x^2/y \log^8 x < m \leqslant ex/X \\ pm=n^2+1 \\ n \leqslant x;\, p \geqslant x}} \log \frac{ex^2}{m}.$$

Therefore, letting $W_x(w)$ be defined by

$$W_x(w) = \sum_{\substack{w < m \leqslant ew \\ pm=n^2+1 \\ n \leqslant x;\, p \geqslant x}} 1,$$

we deduce that

$$T_x''(y) \leqslant \sum_{0 \leqslant \alpha < \log Y} \log (xXe^{\alpha+1}) W_x\Big(\frac{xe^{-\alpha}}{X}\Big),$$

where $Y = (ey \log^8 x)/xX$. Hence, by this, (44), and (45),

$$T_x'(y) \leqslant \sum_{0 \leqslant \alpha < \log Y} \log (xXe^{\alpha+1}) W_x\Big(\frac{xe^{-\alpha}}{X}\Big) + O\Big(\frac{x}{\log x}\Big). \tag{46}$$

In accordance with the principle of the sieve method as outlined in Chapter 1, § 2, the estimations of $V_x(v)$ and $W_x(w)$ depend on explicit formulae for the associated sums that are obtained by replacing p in the conditions of summation by multiples of a given square-free integer λ. The sum corresponding to $V_x(v)$ is of the form

$$\Upsilon(u; \lambda) = \sum_{u < \lambda k \leqslant eu} N_x(\lambda k), \qquad (47)$$

the sum $W_x(w)$ turning out later to be actually of the same type. We must therefore now determine a suitable formula for $\Upsilon(u; \lambda)$. To this end it will be assumed throughout that u and λ satisfy the conditions

$$\left. \begin{array}{c} x^{\frac{4}{5}} < u < x^{\frac{4}{3}} \\[2mm] \lambda \text{ square-free; } \lambda < \min\left(u^{\frac{5}{4}}x^{-1}, xu^{-\frac{3}{4}}\right), \end{array} \right\} \qquad (48)$$

which imply, moreover, that

$$\lambda < x^{\frac{1}{4}} < u^{\frac{1}{2}}. \qquad (49)$$

Here it is useful to note that $u^{\frac{5}{4}}x^{-1} \gtrless xu^{-\frac{3}{4}}$ according as $u \gtrless x$.

3   Transformation of $\Upsilon(u; \lambda)$

We first transform the formula for $N_x(l)$ given by (37). Writing $\psi(t) = [t] - t + \frac{1}{2}$ for any real t, we have

$$N_x(l) = \sum_{\substack{v^2 \equiv -1, \bmod l \\ 0 < v \leqslant l}} \left\{ \frac{x}{l} + \psi\left(\frac{x - v}{l}\right) - \psi\left(-\frac{v}{l}\right) \right\} = \frac{x\rho(l)}{l}$$
$$+ \Psi'_l(x) - \Phi_l, \quad \text{say.} \quad (50)$$

Then, since $\Phi_l = 0$ if $l > 1$, we deduce from this and (47) that

$$\Upsilon(u; \lambda) = \frac{x}{\lambda} \sum_{u < \lambda k \leqslant eu} \frac{\rho(\lambda k)}{k} + \sum_{u < \lambda k \leqslant eu} \Psi'_{\lambda k}(x) = \frac{x}{\lambda} \Sigma_D + \Sigma_E, \quad \text{say.} \quad (51)$$

We transform Σ_E in turn by using the representation of $\psi(t)$ by the Fourier series

$$\psi_{(1)}(t) = \sum_{h=1}^{\infty} \frac{\sin 2\pi h t}{\pi h}.$$

We have the properties

   (i) $\psi(t) = \psi_{(1)}(t)$, unless t is an integer,

  (ii) $\psi_{(1)}(t)$ is boundedly convergent,

 (iii) for $\omega > 1$

$$\sum_{h > \omega} \frac{\sin 2\pi h t}{\pi h} = O\left(\frac{1}{\omega \|t\|}\right),$$

from all of which it follows that

$$\psi(t) = \sum_{1 \leqslant h \leqslant \omega} \frac{\sin 2\pi h t}{\pi h} + O\left\{\min\left(1, \frac{1}{\omega \|t\|}\right)\right\}$$

$$= \psi_\omega(t) + O\{\theta_\omega(t)\}, \quad \text{say.} \quad (52)$$

The Fourier development of $\theta_\omega(t)$ itself is needed. Since $\theta_\omega(t)$ is an even function of t, we have for $\omega > 2$

$$\theta_\omega(t) = \tfrac{1}{2}C_0(\omega) + \sum_{h=1}^{\infty} C_h(\omega) \cos 2\pi h t,$$

where

$$C_h(\omega) = 4 \int_0^{\frac{1}{2}} \theta_\omega(\tau) \cos 2\pi h \tau \, d\tau.$$

Here it is easily verified that

$$C_h(\omega) = \begin{cases} O\left(\dfrac{\log \omega}{\omega}\right), & \text{always,} \\ O\left(\dfrac{\omega}{h^2}\right), & \text{if} \quad h > 0. \end{cases} \quad (53)$$

In order to use the above to put $\Psi_l(x)$ into a suitable form we introduce the important exponential sum

$$\rho(h, l) = \sum_{\substack{\nu^2 \equiv -1, \bmod l \\ 0 < \nu \leqslant l}} e^{2\pi i h \nu / l},$$

where evidently $\rho(h, l)$ is real and $\rho(0, l) = \rho(l)$. By (50) and (52) we have

$$\Psi_l(x) = \Psi_{l, \omega}(x) + O\{\Theta_{l, \omega}(x)\}, \quad (54)$$

where

$$\Psi_{l, \omega}(x) = \sum_{\substack{\nu^2 \equiv -1, \bmod l \\ 0 < \nu \leqslant l}} \psi_\omega\left(\frac{x - \nu}{l}\right), \quad \Theta_{l, \omega}(x) = \sum_{\substack{\nu^2 \equiv -1, \bmod l \\ 0 < \nu \leqslant l}} \theta_\omega\left(\frac{x - \nu}{l}\right).$$

Then, since

$$\psi_\omega\left(\frac{x - \nu}{l}\right) = \frac{1}{\pi} \sum_{1 \leqslant h \leqslant \omega} \frac{1}{h}\left(\sin \frac{2\pi h x}{l} \cos \frac{2\pi h \nu}{l} - \cos \frac{2\pi h x}{l} \sin \frac{2\pi h \nu}{l}\right),$$

we infer that

$$\Psi_{l, \omega}(x) = \frac{1}{\pi} \sum_{1 \leqslant h \leqslant \omega} \frac{1}{h} \rho(h, l) \sin \frac{2\pi h x}{l} \quad (55)$$

on account of the fact that

$$\sum_{\substack{\nu^2 \equiv -1,\ \text{mod } l \\ 0 < \nu \leqslant l}} \sin \frac{2\pi h \nu}{l} = 0.$$

Also, similarly,

$$\Theta_{l,\,\omega}(x) = \tfrac{1}{2} C_0(\omega)\,\rho(l) + \sum_{h=1}^{\infty} C_h(\omega)\,\rho(h, l) \cos \frac{2\pi h x}{l}. \tag{56}$$

Finally, by (51), (54), (55), and (56), we have

$$\Sigma_E = \frac{1}{\pi} \sum_{1 \leqslant h \leqslant \omega} \frac{1}{h} \sum_{u < \lambda k \leqslant eu} \rho(h, \lambda k) \sin \frac{2\pi h x}{\lambda k}$$

$$+ O\!\left(\tfrac{1}{2} C_0(\omega) \sum_{u < \lambda k \leqslant eu} \rho(\lambda k) + \sum_{h=1}^{\infty} C_h(\omega) \sum_{u < \lambda k \leqslant eu} \rho(h, \lambda k) \cos \frac{2\pi h x}{\lambda k} \right), \tag{57}$$

where $\omega = \omega(x, u, \lambda) > 2$.

A further transformation of Σ_E is needed. Let $\mathrm{P}_\lambda^{\pm}(h, u)$ and $\mathrm{P}_\lambda(y)$ be given by

$$\mathrm{P}_\lambda^{\pm}(h, u) = \mathrm{P}_\lambda^{\pm}(h, u, x) = \sum_{u < \lambda k \leqslant eu} \rho(h, \lambda k)\, e^{\pm 2\pi i h x / \lambda k}$$

and

$$\mathrm{P}_\lambda(y) = \sum_{0 < \lambda k \leqslant y} \rho(\lambda k),$$

so that, since $\rho(h, \lambda k)$ is real, we have the properties

$$\left| \mathrm{P}_\lambda^{+}(h, u) \right| = \left| \mathrm{P}_\lambda^{-}(h, u) \right| = S_\lambda(h, u), \quad \text{say,} \tag{58}$$

and

$$S_\lambda(h, u) \leqslant \mathrm{P}_\lambda(eu). \tag{59}$$

Then, by (53), (57), (58), and (59),

$$\Sigma_E = O\!\left(\sum_{1 \leqslant h \leqslant \omega} \frac{S_\lambda(h, u)}{h} \right) + O\!\left(\frac{\log \omega\, \mathrm{P}_\lambda(eu)}{\omega} \right)$$

$$+ O\!\left(\frac{\log \omega}{\omega} \sum_{1 \leqslant h \leqslant \omega} S_\lambda(h, u) \right)$$

$$+ O\!\left(\omega \sum_{\omega < h \leqslant \omega^2} \frac{S_\lambda(h, u)}{h^2} \right) + O\!\left(\omega \mathrm{P}_\lambda(eu) \sum_{h > \omega^2} \frac{1}{h^2} \right)$$

$$= O\!\left(\log \omega \sum_{1 \leqslant h \leqslant \omega} \frac{S_\lambda(h, u)}{h} \right) + O\!\left(\omega \sum_{\omega < h \leqslant \omega^2} \frac{S_\lambda(h, u)}{h^2} \right)$$

$$+ O\!\left(\frac{\log \omega\, \mathrm{P}_\lambda(eu)}{\omega} \right)$$

$$= O\left(\log \omega\, \Sigma_E' \right) + O(\omega \Sigma_E'') + O\!\left(\frac{\log \omega\, \mathrm{P}_\lambda(eu)}{\omega} \right), \quad \text{say.} \tag{60}$$

A formula for Σ_D will be obtained in the next section, while the estimate for Σ_E will flow from the upper bound to be derived later for $S_\lambda(h, u)$.

4   Estimation of Σ_D

We express the Dirichlet series

$$g_\lambda(s) = \sum_{k=1}^{\infty} \frac{\rho(\lambda k)}{k^s}$$

for general square-free numbers λ in terms of the special series $g(s)$ obtained by taking $\lambda = 1$.

Euler's identity gives

$$g_\lambda(s) = \prod_{p\nmid\lambda} \left(\sum_{\alpha=0}^{\infty} \frac{\rho(p^\alpha)}{p^{\alpha s}} \right) \prod_{p|\lambda} \left(\sum_{\alpha=0}^{\infty} \frac{\rho(p^{\alpha+1})}{p^{\alpha s}} \right)$$

for $\sigma > 1$. In this $\rho(p) = \rho(p^\beta)$ for $p \neq 2$ and $\beta > 1$; also $\rho(2) = 1$ but $\rho(2^\beta) = 0$ for $\beta > 1$. Therefore

$$g_\lambda(s) = \prod_{\substack{p\nmid\lambda \\ p\neq 2}} \left(\sum_{\alpha=0}^{\infty} \frac{\rho(p^\alpha)}{p^{\alpha s}} \right) \prod_{p|\lambda} \rho(p) \left(1 - \frac{1}{p^s} \right)^{-1}$$

$$= \rho(\lambda) \prod_{\substack{p\nmid\lambda \\ p\neq 2}} \left(\sum_{\alpha=0}^{\infty} \frac{\rho(p^\alpha)}{p^{\alpha s}} \right) \prod_{p|\lambda} \left(1 - \frac{1}{p^s} \right)^{-1}.$$

Consequently, dividing this equation by the special equation derived from it by letting $\lambda = 1$, we obtain

$$\frac{g_\lambda(s)}{g(s)} = \rho(\lambda) \prod_{p|\lambda} \left(\sum_{\alpha=0}^{\infty} \frac{\rho(p^\alpha)}{p^{\alpha s}} \right)^{-1} \prod_{\substack{p|\lambda \\ p\neq 2}} \left(1 - \frac{1}{p^s} \right)^{-1}.$$

To evaluate this consider the case $\rho(\lambda) > 0$. Here the value of $\rho(p^\alpha)$ for $\alpha > 0$ in the first product is 2 unless $p = 2$ in which event its value is 1 for $\alpha = 1$ and is 0 for $\alpha > 1$. Therefore the first product is

$$\prod_{\substack{p|\lambda \\ p\neq 2}} \left(1 + \frac{2}{p^s} + \frac{2}{p^{2s}} + \dots \right)^{-1} \prod_{\substack{p|\lambda \\ p=2}} \left(1 + \frac{1}{2^s} \right)^{-1}$$

$$= \prod_{p|\lambda} \left(1 + \frac{1}{p^s} \right)^{-1} \prod_{\substack{p|\lambda \\ p\neq 2}} \left(1 - \frac{1}{p^s} \right).$$

Consequently

$$g_\lambda(s) = \rho(\lambda)\,g(s) \prod_{p|\lambda} \left(1 + \frac{1}{p^s}\right)^{-1}, \tag{61}$$

which result still clearly holds when $\rho(\lambda) = 0$.

  Writing

$$\prod_{p|\lambda} \left(1 + \frac{1}{p^s}\right)^{-1} = \sum_{l=1}^{\infty} \frac{a_{l,\lambda}}{l^s}$$

and then using (61), we have

$$\mathrm{P}_\lambda(y) = \rho(\lambda) \sum_{lm \leqslant y/\lambda} a_{l,\lambda}\,\rho(m) = \rho(\lambda) \sum_{l \leqslant y/\lambda} a_{l,\lambda} \sum_{m \leqslant y/\lambda l} \rho(m). \tag{62}$$

The inner sum here is perhaps most easily evaluated by appealing to the theory of quadratic forms. Since the number of sets of primitive representations of m by the form $x^2 + y^2$ of determinant -1 equals the number of roots of the congruence $\nu^2 \equiv -1, \bmod m$, we deduce that $\rho(m)$ is one quarter the number of primitive representations of m by $x^2 + y^2$, there being four representations in each set (Smith [74], arts. 86, 90). Therefore

$$\sum_{m \leqslant z} \rho(m) = \tfrac{1}{4} \sum_{r^2 + s^2 \leqslant z} \sum_{\substack{d|r \\ d|s}} \mu(d) = \tfrac{1}{4} \sum_{d \leqslant z^{\frac{1}{2}}} \mu(d) \sum_{r_1^2 + s_1^2 \leqslant z/d^2} 1$$

$$= \tfrac{1}{4} \sum_{d \leqslant z^{\frac{1}{2}}} \mu(d) \left\{ \frac{\pi z}{d^2} + O\!\left(\frac{z^{\frac{1}{2}}}{d}\right) \right\}$$

by the asymptotic formula for the circle problem (a less elementary version of this formula will be needed in Chapter 6), and so

$$\sum_{m \leqslant z} \rho(m) = \frac{\pi z}{4\zeta(2)} + O\!\left(z \sum_{d > z^{\frac{1}{2}}} \frac{1}{d^2}\right) + O(z^{\frac{1}{2}} \log z) = \frac{3z}{2\pi} + O(z^{\frac{1}{2}}). \tag{63}$$

This and (62) give

$$\mathrm{P}_\lambda(y) = \frac{3\rho(\lambda)\,y}{2\pi\lambda} \sum_{l \leqslant y/\lambda} \frac{a_{l,\lambda}}{l} + O\!\left(\frac{\rho(\lambda)\,y^{\frac{3}{4}}}{\lambda^{\frac{3}{4}}} \sum_{l \leqslant y/\lambda} \frac{|a_{l,\lambda}|}{l^{\frac{3}{4}}}\right)$$

$$= \frac{3\rho(\lambda)\,y}{2\pi\lambda} \prod_{p|\lambda} \left(1 + \frac{1}{p}\right)^{-1} + O\!\left(\frac{\rho(\lambda)\,y}{\lambda} \sum_{l > y/\lambda} \frac{|a_{l,\lambda}|}{l}\right)$$

$$\qquad\qquad + O\!\left(\frac{\rho(\lambda)\,y^{\frac{3}{4}}}{\lambda^{\frac{3}{4}}} \sum_{l \leqslant y/\lambda} \frac{|a_{l,\lambda}|}{l^{\frac{3}{4}}}\right).$$

In this
$$\sum_{l>y/\lambda} \frac{|a_{l,\lambda}|}{l} \leqslant \frac{\lambda^{\frac{1}{4}}}{y^{\frac{1}{4}}} \sum_{l>y/\lambda} \frac{|a_{l,\lambda}|}{l^{\frac{3}{4}}} ;$$

therefore
$$P_\lambda(y) = \frac{3\rho(\lambda)y}{2\pi\lambda} \prod_{p|\lambda} \left(1+\frac{1}{p}\right)^{-1} + O\left(\frac{\rho(\lambda)y^{\frac{3}{4}}}{\lambda^{\frac{3}{4}}} \sum_{l=1}^{\infty} \frac{|a_{l,\lambda}|}{l^{\frac{3}{4}}}\right)$$

$$= \frac{3\rho(\lambda)y}{2\pi\lambda} \prod_{p|\lambda} \left(1+\frac{1}{p}\right)^{-1} + O\left(\frac{\rho(\lambda)\,\sigma_{-\frac{3}{4}}(\lambda)\,y^{\frac{3}{4}}}{\lambda^{\frac{3}{4}}}\right). \qquad (64)$$

We pass from this to the required formula for Σ_D by partial summation. Substituting (64) in the formula

$$\Sigma_D = \lambda \int_u^{eu} \frac{dP_\lambda(t)}{t} = {}^{eu}_{\;u}\left[\frac{\lambda P_\lambda(t)}{t}\right] + \lambda \int_u^{eu} \frac{P_\lambda(t)\,dt}{t^2},$$

we obtain after a short calculation

$$\Sigma_D = \rho(\lambda) \prod_{p|\lambda} \left(1+\frac{1}{p}\right)^{-1} \cdot \frac{3}{2\pi} \int_u^{eu} \frac{dt}{t} + O\left(\frac{\rho(\lambda)\,\sigma_{-\frac{3}{4}}(\lambda)\,\lambda^{\frac{1}{4}}}{u^{\frac{1}{4}}}\right),$$

since the contribution to the term in square brackets due to the explicit part in (64) vanishes. Consequently

$$\Sigma_D = \frac{3\rho(\lambda)}{2\pi} \prod_{p|\lambda} \left(1+\frac{1}{p}\right)^{-1} + O\left(\frac{\lambda^{\frac{1}{4}+\epsilon}}{u^{\frac{1}{4}}}\right). \qquad (65)$$

5   Estimation of $P_\lambda(h, u)$

The method depends on the theory of quadratic forms as outlined, for example, in Smith's Report [74], arts. 86, 90, the application being particularly simple because the determinant involved here is -1 with class number 1. Evidently by (58) it is enough to restrict our attention to the sum $P_\lambda^+(h, u)$, from which for brevity it is now appropriate to leave out the $+$ symbol.

Choose $x^2 + y^2$ as the representative form in the single class of forms of determinant -1. Then, if

$$\lambda k = r^2 + s^2 \qquad (66)$$

is a primitive representation of λk by the form, the root of the congruence $\nu^2 \equiv -1$, mod λk, appertaining to this representation is given by

$$\nu = r\rho + s\sigma,$$

where ρ, σ satisfy $r\sigma - s\rho = 1$. Hence a typical value of $\nu/\lambda k$ in $\rho(h, \lambda k)$ is given by

$$\frac{\nu}{\lambda k} = \frac{r\rho + s\sigma}{r^2 + s^2}.$$

This gives, for $r \neq 0$,

$$\frac{\nu}{\lambda k} = \frac{\rho(r^2 + s^2) + s}{r(r^2 + s^2)} = -\frac{\bar{s}}{r} + \frac{s}{r(r^2 + s^2)}, \tag{67}$$

where $\bar{s}$ is defined, mod r, by $s\bar{s} \equiv 1$, mod r. It gives similarly, for $s \neq 0$,

$$\frac{\nu}{\lambda k} = \frac{\bar{r}}{s} - \frac{r}{s(r^2 + s^2)}, \tag{68}$$

where $r\bar{r} \equiv 1$, mod s. Therefore, defining $\theta_{r,s}$ to be the expression for

$$\frac{\nu}{\lambda k} + \frac{x}{\lambda k}$$

in terms of r, s given by (66) and one of (67) and (68), we have

$$P_\lambda(h, u) = \tfrac{1}{4} \sum_{\substack{u < r^2 + s^2 \leqslant eu \\ r^2 + s^2 \equiv 0, \, \text{mod} \, \lambda \\ (r, s) = 1}} e^{2\pi i h \theta_{r,s}},$$

since there are four representations in each set. The estimation is continued by writing

$$P_\lambda(h, u) = \tfrac{1}{4}\Big(\sum_{|s| < |r|} + \sum_{|r| < |s|} + \sum_{|r| = |s| = 1} \Big) = \tfrac{1}{4}\Sigma_F + \tfrac{1}{4}\Sigma_G + O(1), \tag{69}$$

say, and then considering Σ_F and Σ_G.

The treatment of Σ_F and Σ_G depends on the theory of an exponential sum known as Kloosterman's sum. First introduced by Kloosterman in his researches on the representation of numbers by positive definite quaternary quadratic forms [48], this sum has subsequently been found to be of significance in connection with a variety of interesting problems in the theory of numbers. In such situations the important feature is usually that a good non-trivial bound can be found either for the sum itself or for 'incomplete' sums corresponding to some extra restriction on the variable of summation. Defined as

$$S(a, b; \, k) = \sum_{\substack{0 < h \leqslant k \\ (h, \, k) = 1}} \exp\frac{2\pi i(ah + b\bar{h})}{k},$$

where $\bar{h}$ is given, mod k, by $h\bar{h} \equiv 1$, mod k, the Kloosterman sum has the multiplicative property [29], [30] that for $(k_1, k_2) = 1$

$$S(a, b; k_1 k_2) = S(a, b_1; k_1) S(a, b_2; k_2),$$

where b_1, b_2 are determined, modulis k_1, k_2, respectively, by the congruence

$$b_1 k_2^2 + b_2 k_1^2 \equiv b, \text{ mod } k_1 k_2.$$

It is enough therefore to consider the estimation of the sum in the case where the denominator in the sum is a prime-power.

Deep methods due to Weil [81] are needed to furnish the best possible results when the denominator is a prime (an interesting version of this work has recently been given by Chalk and Smith [6]). These give, for the non-trivial case a, $b \not\equiv 0$, mod p, the inequality

$$|S(a, b; p)| \leqslant 2p^{\frac{1}{2}},$$

from which the universal inequality

$$|S(a, b; p)| \leqslant 2p^{\frac{1}{2}}\{(b, p)\}^{\frac{1}{2}}$$

easily follows by the properties of the Ramanujan sum (for the latter, see [29]).

The case in which the denominator is a prime-power with exponent exceeding 1 usually presents less difficulty, it having been shewn by Salié [66] that an elementary method gives

$$|S(a, b; p^\alpha)| \leqslant 3p^{\frac{1}{2}\alpha}$$

for $\alpha > 1$ and $a, b \not\equiv 0$, mod p. From this and the previous paragraph it is then inferred without difficulty that

$$|S(a, b; p^\alpha)| \leqslant d(p^\alpha) p^{\frac{1}{2}\alpha}\{(b, p^\alpha)\}^{\frac{1}{2}}$$

for any α.

The general inequality

$$|S(a, b; k)| \leqslant d(k) k^{\frac{1}{2}}\{(b, k)\}^{\frac{1}{2}} = O(k^{\frac{1}{2}+\epsilon}\{(b, k)\}^{\frac{1}{2}}) \tag{70}$$

therefore holds in virtue of the multiplicative property of the sum. On this is now based the demonstration of a number of lemmata that lead to an estimate for the exponential sums occurring in our subsequent treatment.

LEMMA 4 *Let l, l_1, r be given integers, where $r \neq 0$, and let $0 \leqslant \xi_2 - \xi_1 \leqslant 2|r|$. Then*

$$\sum_{\substack{\xi_1 \leqslant s \leqslant \xi_2 \\ (r,\,s)=1}} \exp \frac{2\pi i(l_1 s + l\bar{s})}{r} = O(|r|^{\frac{1}{2}+\epsilon}\{(l,r)\}^{\frac{1}{2}}).$$

We notice that r can be taken to be positive by changing the signs of l_1 and l_2 if necessary. With this understanding, the left-hand side of the proposed inequality is equal to

$$\frac{1}{r} \sum_{\substack{0 < s \leqslant r \\ (r,\,s)=1}} \exp \frac{2\pi i(l_1 s + l\bar{s})}{r} \sum_{\xi_1 \leqslant m \leqslant \xi_2} \sum_{0 < \mu \leqslant r} \exp \frac{2\pi i \mu(m-s)}{r}$$

$$= \frac{1}{r} \sum_{0 < \mu \leqslant r} S(l_1 - \mu, l; r) \sum_{\xi_1 \leqslant m \leqslant \xi_2} \exp \frac{2\pi i \mu m}{r}$$

$$= O(r^{\frac{1}{2}+\epsilon}\{(l,r)\}^{\frac{1}{2}}) + O\left(\frac{r^{\frac{1}{2}+\epsilon}\{(l,r)\}^{\frac{1}{2}}}{r} \sum_{0 < \mu \leqslant \frac{1}{2}r} \frac{r}{\mu}\right),$$

by (70). This gives the lemma.

LEMMA 5 *Lemma 4 still holds if l_1 is any real number.*

This is deduced from Lemma 4 by a method due to Estermann. Let $l_2 = [l_1]$ so that $l_1 = l_2 + l_3$ with $0 \leqslant l_3 < 1$. Then, since

$$\exp \frac{2\pi i(l_1 s + l\bar{s})}{r} = \exp \frac{2\pi i(l_2 s + l\bar{s})}{r} \exp \frac{2\pi i l_3 s}{r},$$

we can use partial summation to derive the result from Lemma 4 with l_2 in place of l_1.

LEMMA 6 *If $r \neq 0$ and $0 \leqslant \xi_2 - \xi_1 \leqslant 2|r|$, then*

$$\sum_{\substack{\xi_1 \leqslant s \leqslant \xi_2 \\ s \equiv v,\,\bmod \lambda \\ (r,\,s)=1}} \exp \frac{2\pi i l\bar{s}}{r} = O(|r|^{\frac{1}{2}+\epsilon}\{(l,r)\}^{\frac{1}{2}}).$$

The left-hand side of the required result is equal to

$$\frac{1}{\lambda} \sum_{\substack{\xi_1 \leqslant s \leqslant \xi_2 \\ (r,\,s)=1}} \exp \frac{2\pi i l\bar{s}}{r} \sum_{0 < \mu \leqslant \lambda} \exp \frac{2\pi i \mu(s-v)}{\lambda}$$

$$= \frac{1}{\lambda} \sum_{0 < \mu \leqslant \lambda} \exp \left(-\frac{2\pi i \mu v}{\lambda}\right) \sum_{\substack{\xi_1 \leqslant s \leqslant \xi_2 \\ (r,\,s)=1}} \exp \frac{2\pi i(\mu r \lambda^{-1} s + l\bar{s})}{r}$$

$$= O\left(|r|^{\frac{1}{2}+\epsilon}\{(l,r)\}^{\frac{1}{2}}\right),$$

on applying Lemma 5 to the inner sum.

We also need the following lemma whose proof is elementary.

Lemma 7   *If $h > 0$ and $y \geqslant 1$, then*

$$\sum_{m \leqslant y} \{(h, m)\}^{\frac{1}{2}} = O(y \sigma_{-\frac{1}{2}}(h)).$$

Here

$$\sum_{m \leqslant y} \{(h, m)\}^{\frac{1}{2}} \leqslant \sum_{d \mid h} d^{\frac{1}{2}} \sum_{\substack{m \leqslant y \\ m \equiv 0, \bmod d}} 1 \leqslant y \sum_{d \mid h} \frac{1}{d^{\frac{1}{2}}} = y \sigma_{-\frac{1}{2}}(h),$$

as proposed.

We direct our attention to Σ_F, where it is now always assumed that $h > 0$. The conditions of summation therein evidently imply that $(u/2)^{\frac{1}{2}} < |r| \leqslant (eu)^{\frac{1}{2}}$ so that

$$\Sigma_F = \sum_{(u/2)^{\frac{1}{2}} < |r| \leqslant (eu)^{\frac{1}{2}}} \sum_{\substack{u < r^2 + s^2 \leqslant eu \\ r^2 + s^2 \equiv 0, \bmod \lambda \\ (r, s) = 1; \ |s| < |r|}} e^{2\pi i h \theta_{r,s}}$$

$$= \sum_{(u/2)^{\frac{1}{2}} < |r| \leqslant (eu)^{\frac{1}{2}}} \Sigma_{F, r}, \quad \text{say.} \quad (71)$$

Since in the inner sum the inequalities $-|r| < s < |r|$ and $u - r^2 < s^2 \leqslant eu - r^2$ must both hold, we see that for given r the values of s range through integer values in one or two intervals contained in the interval $(-|r|, |r|)$. To estimate $\Sigma_{F, r}$ we are thus led to consider the sum

$$\Sigma'_{F, r} = \sum_{\substack{\xi_1(r) \leqslant s \leqslant \xi_2(r) \\ r^2 + s^2 \equiv 0, \bmod \lambda \\ (r, s) = 1}} e^{2\pi i h \theta_{r,s}},$$

where ξ_1, ξ_2 are integers satisfying $-|r| < \xi_1 \leqslant \xi_2 < |r|$.

We have

$$\Sigma'_{F, r} = \sum_{\substack{r^2 + v^2 \equiv 0, \bmod \lambda \\ 0 < v \leqslant \lambda}} \sum_{\substack{\xi_1(r) \leqslant s \leqslant \xi_2(r) \\ s \equiv v, \bmod \lambda \\ (r, s) = 1}} e^{2\pi i h \theta_{r,s}}$$

$$= \sum_{\substack{r^2 + v^2 \equiv 0, \bmod \lambda \\ 0 < v \leqslant \lambda}} \Sigma'_{F, r, v}, \quad \text{say.} \quad (72)$$

Writing

$$\phi_h(r, s) = \exp\left(\frac{2\pi i h s}{r(r^2 + s^2)} + \frac{2\pi i h x}{r^2 + s^2}\right),$$

we infer from (66), (67), and (72) that

$$\Sigma'_{F,r,v} = \sum_{\substack{\xi_1(r) \leqslant s \leqslant \xi_2(r) \\ s \equiv v, \bmod \lambda \\ (r,s)=1}} \exp\left(-\frac{2\pi i h \bar{s}}{r}\right) \phi_h(r,s),$$

which when transformed by partial summation becomes

$$\Sigma'_{F,r,v} = \sum_{\xi_1(r) \leqslant \mu \leqslant \xi_2(r)} g_h(\mu)\{\phi_h(r,\mu) - \phi_h(r,\mu+1)\}$$
$$+ g_h(\{\xi_2(r)\})\,\phi_h(r,\xi_2(r)+1), \quad (73)$$

where

$$g_h(\mu) = \sum_{\substack{\xi_1(r) \leqslant s \leqslant \mu \\ s \equiv v, \bmod \lambda \\ (r,s)=1}} \exp\left(-\frac{2\pi i h \bar{s}}{r}\right).$$

In (73)

$$\phi_h(r,\mu) - \phi_h(r,\mu+1) = O\left(\frac{hx}{|r|^3}\right),$$

since $|\mu| < |r|$. Therefore, by this and Lemma 6,

$$\Sigma'_{F,r,v} = O\left(\frac{x^{1+\epsilon}h\{(h,r)\}^{\frac{1}{2}}}{|r|^{\frac{5}{2}}} \sum_{\xi_1(r) \leqslant \mu \leqslant \xi_2(r)} 1\right) + O(|r|^{\frac{1}{2}+\epsilon}\{(h,r)\}^{\frac{1}{2}})$$
$$= O(x^{1+\epsilon}|r|^{-\frac{3}{2}}h\{(h,r)\}^{\frac{1}{2}}) + O(|r|^{\frac{1}{2}+\epsilon}\{(h,r)\}^{\frac{1}{2}}),$$

from which and (72) we infer the first part of

$$\left.\begin{matrix}\Sigma'_{F,r} \\ \Sigma_{F,r}\end{matrix}\right\} = O(x^{1+\epsilon}|r|^{-\frac{3}{2}}h\{(h,r)\}^{\frac{1}{2}}) + O(|r|^{\frac{1}{2}+\epsilon}\{(h,r)\}^{\frac{1}{2}}), \quad (74)$$

since the congruence $r^2 + v^2 \equiv 0$, $\bmod \lambda$, has at most $d(\lambda)$ roots in v. The second part follows from the first part by the discussion at the end of the previous paragraph.

We can complete the estimation of $P_\lambda(h,u)$. First, by (71) and (74) and then by Lemma 7 and partial summation,

$$\Sigma_F = O(x^{1+\epsilon}u^{-\frac{1}{4}}h\sigma_{-\frac{1}{2}}(h)) + O(u^{\frac{3}{4}+\epsilon}\sigma_{-\frac{1}{2}}(h)).$$

Since we see in turn through (69) that Σ_G and $P_\lambda(h,u)$ satisfy a similar inequality, we have finally that

$$P_\lambda(h,u) = O(x^{1+\epsilon}u^{-\frac{1}{4}}h\sigma_{-\frac{1}{2}}(h)) + O(u^{\frac{3}{4}+\epsilon}\sigma_{-\frac{1}{2}}(h)). \quad (75)$$

6   Estimation of Σ_E and $\Upsilon(u; \lambda)$

The estimate for Σ_E follows from (58), (60), and (75). First we have

$$\Sigma'_E = O\left(x^{1+\epsilon}u^{-\frac{1}{4}} \sum_{1 \leqslant h \leqslant \omega} \sigma_{-\frac{1}{2}}(h)\right) + O\left(u^{\frac{3}{4}+\epsilon} \sum_{1 \leqslant h \leqslant \omega} \frac{\sigma_{-\frac{1}{2}}(h)}{h}\right)$$

$$= O(\omega x^{1+\epsilon}u^{-\frac{1}{4}}) + O(u^{\frac{3}{4}+\epsilon}\log\omega). \tag{76}$$

Next

$$\Sigma''_E = O\left(x^{1+\epsilon}u^{-\frac{1}{4}} \sum_{\omega < h \leqslant \omega^2} \frac{\sigma_{-\frac{1}{2}}(h)}{h}\right) + O\left(u^{\frac{3}{4}+\epsilon} \sum_{h > \omega} \frac{\sigma_{-\frac{1}{2}}(h)}{h^2}\right)$$

$$= O(x^{1+\epsilon}u^{-\frac{1}{4}}\log\omega) + O(\omega^{-1}u^{\frac{3}{4}+\epsilon}), \tag{77}$$

while, by (64),

$$P_\lambda(eu) = O\left(\frac{d(\lambda)\,u}{\lambda}\right). \tag{78}$$

Therefore, by (60), (76), (77), and (78),

$$\Sigma_E = O(\omega\log\omega . x^{1+\epsilon}u^{-\frac{1}{4}}) + O(u^{\frac{3}{4}+\epsilon}\log^2\omega) + O\left(\frac{u^{1+\epsilon}\log\omega}{\omega\lambda}\right).$$

Let ω be chosen so that $\omega x u^{-\frac{1}{4}} = 4u\omega^{-1}\lambda^{-1}$ with the consequence that $\omega = 2u^{\frac{5}{8}}x^{-\frac{1}{2}}\lambda^{-\frac{1}{2}} > 2$ by (48). Substituting this value of ω we obtain finally

$$\Sigma_E = O(x^{\frac{1}{2}+\epsilon}u^{\frac{3}{8}}\lambda^{-\frac{1}{2}}) + O(u^{\frac{3}{4}+\epsilon}) = O(x^{\frac{1}{2}+\epsilon}u^{\frac{3}{8}}\lambda^{-\frac{1}{2}}), \tag{79}$$

since $u^{\frac{3}{4}} < x^{\frac{1}{2}}u^{\frac{3}{8}}\lambda^{-\frac{1}{2}}$ by (48).

The estimate for $\Upsilon(u; \lambda)$ is now immediate. Collecting together the results of (51), (65), and (79), we obtain

$$\Upsilon(u; \lambda) = \frac{3x\rho(\lambda)}{2\pi\lambda} \prod_{p | \lambda} \left(1 + \frac{1}{p}\right)^{-1} + O(x^{\frac{1}{2}+\epsilon}u^{\frac{3}{8}}\lambda^{-\frac{1}{2}}), \tag{80}$$

since (48) implies that the error term due to (65) may be absorbed into that due to (79).

7   Application of the sieve method

A suitable formula for $\Upsilon(u; \lambda)$ having been obtained, Selberg's upper bound sieve method is applied to the estimation of $V_x(v)$ and $W_x(w)$.

In the estimation of $V_x(v)$ it will now be assumed that $x \leqslant v < x^{\frac{12}{11}}$ so that in particular v satisfies the first condition imposed on u by (48). Let d denote a square-free number (possibly 1) composed entirely of prime factors p such that $p = 2$ or $p \equiv 1, \mathrm{mod}\, 4$, and let $\lambda_d = \lambda_{d,z}$ be taken as at the beginning of § 2, Chapter 1, with $1 < z < x^{\frac{1}{2}} v^{-\frac{3}{8}}$. Then, as $v > z$, we have

$$V_x(v) \leqslant \sum_{v < l \leqslant ev} N_x(l)(\sum_{d \mid l} \lambda_d)^2 = \sum_{d_1, d_2 \leqslant z} \lambda_{d_1} \lambda_{d_2} \sum_{\substack{v < l \leqslant ev \\ l \equiv 0,\, \mathrm{mod}\, [d_1, d_2]}} N_x(l)$$

$$= \sum_{d_1, d_2 \leqslant z} \lambda_{d_1} \lambda_{d_2} \Upsilon(v; [d_1, d_2]).$$

Therefore, since here $[d_1, d_2] < xv^{-\frac{3}{4}}$, we have

$$V_x(v) \leqslant \frac{3x}{2\pi} \sum_{d_1,\, d_2 \leqslant z} \frac{\lambda_{d_1} \lambda_{d_2}}{f([d_1, d_2])} + O\left(x^{\frac{1}{2} + \epsilon} v^{\frac{3}{8}} \sum_{d_1,\, d_2 \leqslant z} \frac{|\lambda_{d_1}|\, |\lambda_{d_2}|}{([d_1, d_2])^{\frac{1}{2}}}\right) \quad (81)$$

by (80), where $f(n)$ is the multiplicative function defined by

$$\frac{1}{f(n)} = \frac{\rho(n)}{n} \prod_{p \mid n} \left(1 + \frac{1}{p}\right)^{-1}.$$

The first term in (81) is to be minimized as in Chapter 1. Setting

$$f_1(d) = f(d) \prod_{p \mid d} \left(1 - \frac{1}{f(p)}\right),$$

we have

$$V_x(v) \leqslant \frac{3x}{2\pi \sum_{d \leqslant z} \dfrac{1}{f_1(d)}} + O\left(x^{\frac{1}{2} + \epsilon} v^{\frac{3}{8}} \sum_{d_1,\, d_2 \leqslant z} \frac{1}{([d_1, d_2])^{\frac{1}{2}}}\right), \quad (82)$$

since $\lambda_d = O(1)$ by (18).

To evaluate the first term in the right-hand side of (82) we modify the procedure that was mentioned in connection with (19). Let now d' indicate, generally, either 1 or any number (square-free or otherwise) whose prime factors divide d (i.e. $p \mid d'$ implies $p \mid d$). If $p = 2$, then

$$\frac{1}{f_1(p)} = \frac{1}{p};$$

if $p \equiv 1, \mathrm{mod}\, 4$, then

$$f_1(p) = \frac{p}{2}\left(1 + \frac{1}{p}\right) - 1 = \frac{p}{2}\left(1 - \frac{1}{p}\right)$$

so that
$$\frac{1}{f_1(p)} = \frac{2}{p}\left(1 - \frac{1}{p}\right)^{-1}.$$

Therefore, as $f_1(d)$ is multiplicative, we have
$$\frac{1}{f_1(d)} = \frac{\rho(d)}{d} \prod_{\substack{p|d \\ p\neq 2}}\left(1 - \frac{1}{p}\right)^{-1} = \sum_{d'}\frac{\rho(dd')}{dd'},$$

using the fact that $\rho(2^\alpha) = 0$ if $\alpha > 1$. We deduce that for $y > y(\eta_1)$
$$\sum_{d\leqslant y}\frac{1}{f_1(d)} = \sum_{d\leqslant y}\sum_{d'}\frac{\rho(dd')}{dd'} \geqslant \sum_{m\leqslant y}\frac{\rho(m)}{m} \geqslant \frac{3(1-\eta_1)\log y}{2\pi} \quad (83)$$

by (63) and partial summation. Also the error term in (81) is
$$O\left(x^{\frac{1}{2}+\epsilon}v^{\frac{3}{8}}\sum_{d\leqslant z}\sum_{\substack{l_1,\, l_2\leqslant z/d \\ (l_1,\, l_2)=1}}\frac{1}{d^{\frac{1}{2}}l_1^{\frac{1}{2}}l_2^{\frac{1}{2}}}\right) = O\left(x^{\frac{1}{2}+\epsilon}v^{\frac{3}{8}}\sum_{d\leqslant z}\frac{z}{d^{\frac{3}{2}}}\right) = O(x^{\frac{1}{2}+\epsilon}v^{\frac{3}{8}}z). \quad (84)$$

Choosing z to be $x^{\frac{1}{2}-\eta}v^{-\frac{3}{8}}$ (consistent with earlier conditions), we infer the required estimate
$$V_x(v) < \frac{(1+\eta_2)x}{\log(x^{\frac{1}{2}}v^{-\frac{3}{8}})} \quad (85)$$

from (82), (83), and (84).

The estimation of $W_x(w)$ being very similar to that of $V_x(v)$ after the first stage, we suppress all but the earlier details. Here we assume $x^{\frac{4}{5}} < w < x$ and $1 < z < x^{\frac{2}{7}}w^{-\frac{3}{14}}$. Then, since $z < x$ (this relating to $p \geqslant x$ in the definition of $W_x(w)$),
$$W_x(w) \leqslant \sum_{\substack{w<m\leqslant ew \\ lm=n^2+1 \\ n\leqslant x}}\left(\sum_{d|l}\lambda_d\right)^2 = \sum_{d_1,\, d_2\leqslant z}\lambda_{d_1}\lambda_{d_2}\sum_{\substack{w<m\leqslant ew \\ m[d_1,\, d_2]\, r=n^2+1 \\ n\leqslant x}}1$$
$$= \sum_{d_1,\, d_2\leqslant z}\lambda_{d_1}\lambda_{d_2}\Upsilon([d_1,d_2]w;[d_1,d_2]).$$

The conditions on w and z shew that in this sum $u = [d_1,d_2]w$ and $\lambda = [d_1,d_2]$ satisfy (48). Therefore, by (80), we obtain as an analogue of (81)
$$W_x(w) \leqslant \frac{3x}{2\pi}\sum_{d_1,\, d_2\leqslant z}\frac{\lambda_{d_1}\lambda_{d_2}}{f([d_1,d_2])} + O\left(x^{\frac{1}{2}+\epsilon}w^{\frac{3}{8}}\sum_{d_1,\, d_2\leqslant z}\frac{|\lambda_{d_1}|\,|\lambda_{d_2}|}{([d_1,d_2])^{\frac{1}{8}}}\right)$$
$$= \frac{3x}{2\pi}\sum_{d_1,\, d_2\leqslant z}\frac{\lambda_{d_1}\lambda_{d_2}}{f([d_1,d_2])} + O(x^{\frac{1}{2}+\epsilon}w^{\frac{3}{8}}z^{\frac{7}{4}}) < \frac{(1+\eta_2)x}{\log(x^{\frac{2}{7}}w^{-\frac{3}{14}})}, \quad (86)$$

on choosing z to be $x^{\frac{2}{7}-\eta}w^{-\frac{3}{14}}$.

8   The greatest prime divisor

We arrive at our final result by using the estimates for $V_x(v)$ and $W_x(w)$ to obtain an upper bound for $T_x(y)$ for $y = x^{\frac{11}{10}}$. To this end let $\gamma = \log x$.

First it is necessary to estimate $T_x(xX)$. Since $xe^\alpha < x^{\frac{12}{11}}$ in (43), we have by (85)

$$T_x(xX) < x(1+\eta_2) \sum_{0 \leqslant \alpha < \log X} \frac{\alpha + \gamma + 1}{\frac{1}{8}\gamma - \frac{3}{8}\alpha}$$

$$< 8x(1+\eta_2) \int_0^{\log X} \frac{(\gamma + t)\,dt}{\gamma - 3t} + O(x)$$

$$= 8x(1+\eta_2) \left[-\tfrac{1}{3}t - \frac{4\gamma \log(\gamma - 3t)}{9} \right]_0^{\log X} + O(x)$$

$$= 8x(1+\eta_2)(-\tfrac{1}{33} + \tfrac{4}{9}\log \tfrac{11}{8})\log x + O(x)$$

$$= (\cdot 89010\ldots)(1+\eta_2)\,x\log x + O(x) < \cdot 8902 x \log x. \quad (87)$$

Next, for $y = x^{\frac{11}{10}}$, equation (86) implies that

$$T'_x(y) < x(1+\eta_2) \sum_{0 \leqslant \alpha < \log Y} \frac{\alpha + \frac{12}{11}\gamma + 1}{\frac{1}{11}\gamma + \frac{3}{14}\alpha} + O\left(\frac{x}{\log x}\right),$$

since then $xX^{-1}e^{-\alpha} > x^{\frac{4}{5}}$ in (46). Therefore in this case

$$T'_x(y) < 14x(1+\eta_2) \int_0^{\log Y} \frac{(12\gamma + 11t)\,dt}{14\gamma + 33t} + O(x)$$

$$= 14x(1+\eta_2) \left[\tfrac{1}{3}t + \frac{2\gamma \log(14\gamma + 33t)}{9} \right]_0^{\log Y} + O(x)$$

$$= 14x(1+\eta_2)(\tfrac{1}{330} + \tfrac{2}{9}\log \tfrac{143}{140})\log x + O(x \log \log x)$$

$$= (\cdot 10808\ldots)(1+\eta_2)\,x\log x + O(x \log \log x)$$

$$< \cdot 1081 x \log x. \quad (88)$$

Combining (87) and (88) we thus have from (42) that

$$T_x(x^{\frac{11}{10}}) < \cdot 9983 x \log x.$$

If this is compared with (41), we deduce at once that $P_x > x^{\frac{11}{10}}$ ($x > x_0$). We thus have

THEOREM 1   *The greatest prime factor of*

$$\prod_{n \leqslant x} (n^2 + 1)$$

exceeds $x^{\frac{11}{10}}$ for all sufficiently large values of x.

9   Other applications of the method

The above method was later closely followed by Ramachandra [63] in his investigation of the greatest prime factor of the integers in an interval. In this treatment the counterparts of the exponential sums used here are those that occur in van der Corput's theory (see [77], Chapter 5).

Other investigations on greatest prime factors of numbers in sequences have also depended to some extent on the ideas of this chapter. For instance, Greaves [24] has proved that, if P_x is the greatest prime factor of

$$\prod_{0 < |f| \leqslant x} f(u, v),$$

where $f(u, v)$ is an irreducible binary cubic form, then

$$P_x > x^{\frac{2}{3} + \alpha}$$

for some positive constant α and $x > x_0$. In much the same spirit Motohashi [59] shewed in 1970 that, for any non-zero integer a, the greatest prime factor P_x of

$$\prod_{-a < p \leqslant x} (p + a)$$

satisfies

$$P_x > x^\theta,$$

where $\theta = \cdot 611059\ldots$ The interest of this problem of course rests largely on its connection with the prime twin conjecture just as Chebyshev's problem is connected with conjectures about the representation of primes by quadratic polynomials. However, the author's work on the Brun–Titchmarsh theorem on average, as outlined briefly in Chapter 1, § 4, leads to the slightly stronger value $\cdot 6197\ldots$ for the exponent θ [40]. Recently the author further improved this result by shewing that θ could be taken to be any number less than $\frac{5}{8}$ [42]; this was effected by replacing the use of the Brun–Titchmarsh theorem by a sieve method that was related to the distribution of numbers of the form lq in arithmetic progressions, q here denoting a prime.

Returning now to the Chebyshev problem itself, we remark that the method is successful because it is possible to shew that the roots of the congruences $\nu^2 \equiv -1$, $\mathrm{mod}\, l$, are very evenly distributed in the sense that the ratios ν/l are uniformly distributed, mod 1, when ordered suitably as a sequence according to increasing values of l. If results of comparable power could be established for the roots of polynomial congruences of higher degree, then many other interesting consequences would ensue. In particular the result

$$P_x > x^{1+\alpha_r} \tag{89}$$

could be proved for a polynomial of degree r, and substantial improvements could be made to our later results in Chapter 4 about the power-free values taken by polynomials.

The author has indeed been able to shew by an alternative method that the roots ν of the congruence $f(\nu) \equiv 0$, $\mathrm{mod}\, l$, are evenly distributed in the sense explained above provided $f(n)$ is irreducible [33]. The estimates implicit in the method are not, however, sufficiently sharp for this result to be of any substantial help in the resolution of the more significant relevant problems.

There is one possible approach to this question of uniform distribution that shews some promise. This is to devise a method for explicitly constructing all the roots of $f(\nu) \equiv 0$, $\mathrm{mod}\, l$, for $l \leqslant M$ by means of some appropriate generalization of the theory that connects the roots of $\nu^2 \equiv D$, $\mathrm{mod}\, l$, with the representation of l by quadratic forms of determinant D. Such a generalization actually exists, although no mention of it appears to have been given in the literature. Choosing for simplicity the cubic congruence

$$\nu^3 \equiv 2, \mathrm{mod}\, l,$$

we take a corresponding cubic factorizable form associated with the pure cubic extension $Q(\sqrt[3]{2})$ of the rational field Q. Since the class number is in this case 1, we can take this form to be

$$\phi(x, y, z) = x^3 + 2y^3 + 4z^3 - 6xyz.$$

Next we say that the representation of l by $\phi(x, y, z)$ is *primitive* provided

$$(X, Y, Z) = 1,$$

where

$$X = x^2 - 2yz, \quad Y = y^2 - zx, \quad Z = 2z^2 - xy.$$

Then it is possible to shew that each such primitive representation gives rise to a unique root, mod l, of $\nu^3 \equiv 2$, mod l; conversely, each root gives rise to a *set* of primitive representations that are associated with each other through the automorphic transformations of the form. Furthermore this correspondence enables the set of ratios ν/l to be expressed explicitly in terms of x, y, z, where x, y, z are restricted to lie in a certain frustum in order that each root be obtained just once. The expression then achieved for

$$\sum_{l \leqslant M} \sum_{\substack{\nu^3 \equiv 2, \, \text{mod} \, l \\ 0 < \nu \leqslant l}} e^{2\pi i h \nu / l}$$

is complicated, but it is possible after some transformations to make it depend on incomplete Kloosterman sums. Unfortunately, the number of terms in these sums is too small compared with the moduli for Lemma 4 to yield non-trivial results. Yet the analysis can be pushed through to a successful conclusion provided the natural conjecture about the true size of incomplete Kloosterman sums of short length be assumed. A conditional proof of (89) is thus possible for the cubic $f(n) = n^3 - 2$.

3. On Artin's conjecture – an application of the simple asymptotic sieve

1 Introduction

In 1927 Artin enunciated the hypothesis, now usually known as Artin's conjecture, that for any given integer a other than 1, -1, or a perfect square there exist infinitely many primes p for which a is a primitive root, modulo p. Furthermore, letting $N_a(x)$ denote the number of such primes up to the limit x, he was led to conjecture an asymptotic formula of the type

$$N_a(x) \sim \frac{A(a)\,x}{\log x} \quad (x \to \infty)$$

with $A(a) > 0$. Later, however, his suggested value of the constant $A(a)$ was brought into question for certain values of a, since the work of D. H. Lehmer revealed that the formula did not appear to predict values for $N_a(x)$ that were in accord with the numerical evidence. In the light of this knowledge, a revised form of $A(a)$ was then proposed by Heilbronn (see Miller [55]) and others.

In a paper published in 1967 the author proved both these conjectures subject to the assumption that the Riemann hypothesis hold for Dedekind zeta functions over certain Kummerian fields [36]. The value obtained for $A(a)$ in the asymptotic formula was, moreover, in agreement with that suggested by Heilbronn. In a footnote to this paper it was explained that it was already possible to achieve these results on weaker hypotheses, the form of hypothesis adopted in the account having been chosen for the sake of clarity. Although nothing specific in this matter was mentioned, these remarks referred to the discovery that any hypothesis required need only be assumed in respe t of zeta functions over fields of prime degree q of the type $Q(\sqrt[q]{a})$. Subsequently A. I. Vinogradov [80] developed an alternative conditional treatment that was tantamount to a reinterpretation of the author's method in terms of class field theory and the theory of Artin's L-functions.

No further significant progress has been made with this central problem in primitive root theory, even though several interesting papers with some bearing on the matter have appeared. Yet it should be added that the analogues of these conjectures for function fields were demonstrated by Bilharz [1] in 1937 on the supposition, subsequently shewn to be correct by Weil, that the Riemann hypothesis hold for congruence zeta functions.

The principle underlying the author's treatment of the Artin conjectures was that of the simple asymptotic sieve. In this chapter we propose initially to describe the method by giving the conditional proof for the special case $a = 2$. The special choice of a has been made with the intention of making the exposition as short and simple as possible, and it should be emphasized that the general case differs from ours only in that there are some extra points of detail that require discussion. Here, as in the original paper, we have chosen the hypothesis in a form that makes for clarity in presentation.

In the latter part of the chapter we then pass on to the important question of how we may relax the dependence of our results on unproved hypotheses.

We shall refer to Artin's conjecture again in Chapter 7 when discussing the occurrence of prime numbers in a particular sparse sequence.

2 Formulation of the method

A simple principle, readily inferred from the theory of indices, serves to distinguish the prime numbers p for which 2 is a primitive root, modulo p. The index calculus shews that for $p \neq 2$ and any prime divisor q of $p - 1$ the solubility of the congruence

$$\nu^q \equiv 2, \bmod p,$$

is equivalent to the divisibility of the index of 2 (to any base) by q. We thus have the criterion that 2 is a primitive root, modulo p, if, and only if, $p \neq 2$ and there is no prime divisor q of $p - 1$ for which 2 is a qth power residue, mod p.

We introduce the notation for utilizing this principle through the simple asymptotic sieve. First it is assumed throughout that $p > 2$. Next, for any such prime p and for any prime q, let the simultaneous conditions

$$2 \text{ is a } q\text{th power residue, mod} p; \quad q|(p-1)$$

be indicated by $R(q, p)$; let also, for any square-free number k, the generalized symbol $R(k, p)$ indicate the conjunction of $R(q, p)$ for all prime divisors q of k.

We also require a notation for certain sums that count the number of primes p not exceeding x with a specific property. First $N(x, \eta) = N_2(x, \eta)$ appertains to the property that p do not satisfy $R(q, p)$ for any prime q not exceeding η, where the criterion for primitive roots implies that $N_2(x) = N(x, x-1)$ because $R(q, p)$ cannot hold for $q > p - 1$; next the sum $P(x, k)$ for any square-free number k appertains to the property that $R(k, p)$ hold (no condition being implied if $k = 1$); finally $M(x, \eta_1, \eta_2)$ appertains to the property that $R(q, p)$ hold for at least one prime q satisfying $\eta_1 < q \leqslant \eta_2$.

We infer from these definitions and from the principles of Chapter 1, §1, that

$$N_2(x) = N(x, \xi_1) + O\{M(x, \xi_1, \xi_2)\} + O\{M(x, \xi_2, \xi_3)\} \\ + O\{M(x, \xi_3, x-1)\}, \quad (90)$$

where $\quad \xi_1 = \tfrac{1}{6} \log x, \quad \xi_2 = x^{\frac{1}{2}} \log^{-2} x, \quad \xi_3 = x^{\frac{1}{2}} \log x.$

We consider the last two terms in this at once since they can be estimated fairly easily without any hypothesis.

To estimate the first we use the inequality

$$M(x, \xi_2, \xi_3) \leqslant \sum_{\xi_2 < q \leqslant \xi_3} P(x, q).$$

Retaining only the condition $q|(p-1)$ in $R(q, p)$, we also have

$$P(x, q) \leqslant \sum_{\substack{p \leqslant x \\ p \equiv 1, \text{mod } q}} 1 < \frac{A_1 x}{(q-1) \log (x/q)}$$

by Lemma 1. We therefore deduce that

$$M(x, \xi_2, \xi_3) = O\left(\frac{x}{\log x} \sum_{\xi_2 < q \leqslant \xi_3} \frac{1}{q}\right) = O\left(\frac{x}{\log^2 x} \sum_{\xi_2 < q \leqslant \xi_3} \frac{\log q}{q}\right),$$

from which it follows by the easier Mertens formula that

$$M(x, \xi_2, \xi_3) = O\left\{\frac{x}{\log^2 x}\left(\log\frac{\xi_3}{\xi_2} + O(1)\right)\right\} = O\left(\frac{x \log\log x}{\log^2 x}\right). \quad (91)$$

Turning our attention to the final term we observe that the condition $R(q, p)$ implies that

$$2^{\frac{p-1}{q}} \equiv 1, \bmod p.$$

Therefore, since $q > x^{\frac{1}{2}}\log x$ and $p \leqslant x$, any primes p that the sum $M(x, \xi_3, x-1)$ counts must divide the positive product

$$\prod_{m < x^{\frac{1}{2}}\log^{-1} x} (2^m - 1).$$

Therefore

$$2^{M(x, \xi_3, x-1)} < \prod_{m < x^{\frac{1}{2}}\log^{-1} x} 2^m,$$

and so

$$M(x, \xi_3, x-1) < \sum_{m < x^{\frac{1}{2}}\log^{-1} x} m = O\left(\frac{x}{\log^2 x}\right). \quad (92)$$

We thus gather from (91) and (92) that (90) may now be expressed in the simplified form

$$N_2(x) = N(x, \xi_1) + O\{M(x, \xi_1, \xi_2)\} + O\left(\frac{x \log\log x}{\log^2 x}\right). \quad (93)$$

We terminate the section by expressing the first two terms in the right-hand side of this equation in terms of $P(x, k)$. First, letting l' indicate either 1 or positive square-free numbers composed entirely of prime factors q not exceeding ξ_1, we have

$$N(x, \xi_1) = \sum_{l'} \mu(l') \, P(x, l'). \quad (94)$$

We note here the upper bound

$$l' \leqslant \prod_{q \leqslant \xi_1} q = e^{\vartheta(\xi_1)} \leqslant e^{2\xi_1} = x^{\frac{1}{3}} \quad (95)$$

that will be required for the subsequent application of (94). Secondly we have

$$M(x, \xi_1, \xi_2) \leqslant \sum_{\xi_1 < q \leqslant \xi_2} P(x, q). \quad (96)$$

3   Application of algebraic number theory

We shew that the primes counted in the sum $P(x, k)$ can be characterized in terms of conditions formulated in the language of algebraic number theory.

The primes contributing to $P(x, k)$ are just those primes p for which the simultaneous conditions

$$\nu^q \equiv 2, \bmod p, \text{ soluble}; \; p \equiv 1, \bmod q,$$

hold for every prime divisor q of k. We therefore infer that the primes in question are identified by the simpler equivalent simultaneous conditions

$$\nu^k \equiv 2, \bmod p, \text{ soluble}; \; p \equiv 1, \bmod k. \qquad (97)$$

To reformulate the above principle in the notation of algebraic number theory we write the rational number field as Q, and indicate, for any algebraic extension M of a field L, the degree of M over L by $[M:L]$. Then the field

$$F = Q(\sqrt[k]{2})$$

generated by a (positive) root of the irreducible equation

$$u^k - 2 = 0$$

has degree k, the discriminant of the defining polynomial having modulus $2^{k-1}k^k$. Similarly, if $\sqrt[k]{1}$ indicates a primitive kth root of unity, then the cyclotomic field

$$Z = Q(\sqrt[k]{1})$$

has degree $\phi(k)$, where the discriminant of the corresponding cyclotomic polynomial is formed entirely from the prime divisors of k.

Now the solubility of $\nu^k \equiv 2$, $\bmod p$, and the condition $p \equiv 1$, $\bmod k$, are together equivalent to the requirement that $\nu^k \equiv 2$, $\bmod p$, have exactly k incongruent roots. Hence, by a famous principle due to Dedekind [9], these conditions are also equivalent to the requirement that $p \nmid k$ and that p factorize in F as a product of k distinct linear prime ideals. Similarly the condition $p \equiv 1$, $\bmod k$, is equivalent to the condition that $p \nmid k$ and p

factorize in Z as a product of $\phi(k)$ distinct linear prime ideals. We thus conclude that (97) is equivalent to the condition that $p \nmid k$ and that p factorize totally in the Kummerian field

$$G = Q(\sqrt[k]{2}, \sqrt[k]{1})$$

as a product of distinct linear prime ideals.

To apply this principle it is requisite to discover the degree

$$n = n(k) = [G\!:\!Q] \tag{98}$$

of G over Q. Observing that it suffices to determine $[G\!:\!Z]$ because

$$[G\!:\!Q] = [G\!:\!Z][Z\!:\!Q] = \phi(k)[G\!:\!Z], \tag{99}$$

we have

$$[G\!:\!Z]\,|\,k$$

in virtue of the fact that Z is Galoisian. Let

$$k = m[G\!:\!Z]. \tag{100}$$

Then, for any prime factor q of m, we have that $[Z(\sqrt[q]{2})\!:\!Z]$ is either 1 or q, and that also $[Z(\sqrt[q]{2})\!:\!Z]\,|\,(k/m)$. As $(k/m,q)=1$ (k square-free), it follows that $[Z(\sqrt[q]{2})\!:\!Z]=1$ and that therefore $\sqrt[q]{2}$ is contained in Z. Since all sub-fields of the Abelian field Z are themselves Abelian, we infer that q cannot be odd. Also q cannot be even either, since it follows from the theory of cyclotomy that the only quadratic sub-fields of Z are of the form $Q(\sqrt{(\pm D)})$ where D is an odd divisor of k. We deduce that $m = 1$ and hence from (98), (99), and (100) that

$$n(k) = k\phi(k). \tag{101}$$

The sum $\mathrm{P}(x, k)$ can now be prepared for the investigation of the next section by expressing it in terms of $\pi(x, k)$, the number of prime ideals $\mathfrak{p}$ in G such that $N\mathfrak{p} \leqslant x$.

We write $\pi(x, k)$ as

$$\pi(x, k) = \pi^{(1)}(x, k) + \pi^{(2)}(x, k), \tag{102}$$

where $\pi^{(1)}(x, k)$ is the contribution to $\pi(x, k)$ due to linear prime ideals that do not divide $2k$, and $\pi^{(2)}(x, k)$ is the remaining contribution. Now, since in G each rational prime p relatively prime to $2k$ either has $n(k)$ linear prime ideal factors or has no such factors, we see that

$$\pi^{(1)}(x, k) = n(k)\,\mathrm{P}(x, k) \tag{103}$$

3

in view of the criterion obtained above for the primes counted by $P(x, k)$. Also

$$\pi^{(2)}(x, k) \leqslant n(k)\,\omega(2k) + n(k) \sum_{p^2 \leqslant x} 1. \tag{104}$$

The required relation

$$k\phi(k)\,P(x, k) = \pi(x, k) + O\{k\phi(k)\,\omega(k)\} + O\{k\phi(k)\,x^{\frac{1}{2}}\} \tag{105}$$

then follows from (101), (102), (103), and (104).

4   Estimation of $\pi(x, k)$ and $P(x, k)$

In this section the theory of Dedekind's zeta function†

$$\zeta(s) = \zeta_k(s) = \sum_{\mathfrak{p}} \frac{1}{(N\mathfrak{p})^s}$$

is applied to the estimation of $\pi(x, k)$, the purpose being in effect to obtain a suitable conditional variant of the classical prime ideal theorem for fields of the form G. Some preliminary remarks and results are necessary, and the reader is referred where appropriate to Landau's book on algebraic number theory [50].

If $\Delta = \Delta(k)$ is the discriminant of G, then the formulae for the discriminants of the fields $Q(\sqrt[q]{1})$ and $Q(\sqrt[q]{2})$ imply the inequality

$$||\Delta| \leqslant (\prod_{q|k} q \prod_{q|k} 2q)^{k\phi(k)} \leqslant (2^{\omega(k)}k^2)^{k\phi(k)} \leqslant k^{A_2 k\phi(k)} = k^{A_2 n}, \tag{106}$$

this evidently remaining true when $k = 1$.

The numbers $r_1 = r_1(k)$, $r_2 = r_2(k)$, $r = r_1 + r_2 - 1$ have their customary meaning; that is, if $G = Q(\theta)$, then r_1 is the number of real conjugates of θ over Q and $2r_2$ is the number of non-real conjugates over Q. It is useful to note that $r_1 = 1$, $r_2 = r = 0$ when $k = 1$, and that generally

$$0 \leqslant r_1, r_2, r \leqslant n(k).$$

The basic properties of $\zeta(s)$ required are:

(i) $\zeta(s)$ is an analytic function regular in every finite part of the plane except for a simple pole at $s = 1$ (Satz 154).

---

† The notation $\zeta(s)$ will be used to denote $\zeta_k(s)$ until the end of § 6, after which it will revert to its usual meaning of the Riemann zeta function.

(ii) $\zeta(s)$ satisfies the functional equation

$$B^s\Gamma^{r_1}(\tfrac{1}{2}s)\,\Gamma^{r_2}(s)\,\zeta(s) = B^{1-s}\Gamma^{r_1}(\tfrac{1}{2}-\tfrac{1}{2}s)\,\Gamma^{r_2}(1-s)\,\zeta(1-s),$$

where $B = B_k = 2^{-r_2}\pi^{-\frac{1}{2}n}\sqrt{|\Delta|}$ (Satz 154).

(iii) $\zeta(s)$ is of finite order in every half-plane $\sigma \geqslant \sigma_1$. More precisely, for every σ_1, there exists a number $u = u(\sigma_1, k)$ such that, as $|t| \to \infty$,

$$t^{-u}\zeta(s) \to 0$$

uniformly with respect to σ (but not necessarily with respect to k) for $\sigma \geqslant \sigma_1$ (Satz 171).

As is customary in the analytic method we first introduce the sum

$$\psi(x) = \psi(x, k) = \sum_{(N\mathfrak{p})^a \leqslant x} \log N\mathfrak{p},$$

since this can be more directly related than $\pi(x, k)$ to the properties of $\zeta(s)$ and its zeros. Denoting a general 'complex' zero of $\zeta(s)$ by $\rho = \beta + i\gamma$, we have

(iv) For non-integral values of x such that $x \geqslant \tfrac{3}{2}$,

$$\psi(x) = x - (r\log x + a_0) - \tfrac{1}{2}r_1\log\left(1 - \frac{1}{x^2}\right) - r_2\log\left(1 - \frac{1}{x}\right) - \sum_\rho \frac{x^\rho}{\rho},$$

where $a_0 = a_0(k)$, the series on the right converging uniformly in any closed interval that does not include an integer (Sätze 197 and 199). Since $\zeta(s)$ is real for real values of s, the values of ρ in this formula occur in conjugate pairs.

First we obtain an upper bound, holding uniformly with respect to k, for $|(s-1)\,\zeta(s)|$ in the strip $-\tfrac{3}{2} \leqslant \sigma \leqslant \tfrac{11}{2}$. For $\sigma \geqslant 2$, we have

$$|\zeta(s)| \leqslant \zeta(2) = \prod_{\mathfrak{p}}\left(1 - \frac{1}{(N\mathfrak{p})^2}\right)^{-1} \leqslant \prod_{p}\left(1 - \frac{1}{p^2}\right)^{-n} = \left(\frac{\pi^2}{6}\right)^n = A_3^n.$$

$$(107)$$

Also, if $\sigma = -\tfrac{3}{2}$, we have from (ii) that

$$|\zeta(s)| = |B|^4 \left|\frac{\Gamma(\tfrac{1}{2}-\tfrac{1}{2}s)}{\Gamma(\tfrac{1}{2}s)}\right|^{r_1}\left|\frac{\Gamma(1-s)}{\Gamma(s)}\right|^{r_2}|\zeta(1-s)|$$

$$\leqslant |B|^4 A_4^{r_1}(|t|+2)^{2r_1} A_4^{r_2}(|t|+2)^{4r_2}|\zeta(1-s)|$$

on using Stirling's formula. We deduce from this, (ii), (107), and the definitions of r_1, r_2 that

$$|\zeta(s)| \leqslant |\Delta|^2(|t|+2)^{2n} A_3^n A_4^n,$$

and then from (106) that

$$|(s-1)\,\zeta(s)| \leqslant \{k(|t|+2)\}^{A_5\,n}. \tag{108}$$

Since by (107) this inequality also holds for $\sigma = \frac{11}{2}$ provided A_5 is chosen appropriately, we have that on both the lines $\sigma = \frac{11}{2}$, $\sigma = -\frac{3}{2}$

$$|Z(s)| \leqslant (A_6 k)^{A_7 n}, \tag{109}$$

where $A_7 = |A_5| + 1$ and

$$Z(s) = \frac{(s-1)\,\zeta(s)}{(s+2)^{A_7 n}}.$$

The function $Z(s)$ is, however, regular for $-\frac{3}{2} \leqslant \sigma \leqslant \frac{11}{2}$ and of finite order. Therefore, by the Phragmén–Lindelöf principle (see Hardy and Riesz [26], Chapter 3), the inequality (109) subsists throughout the strip $-\frac{3}{2} \leqslant \sigma \leqslant \frac{11}{2}$. This implies the validity of the required inequality

$$|(s-1)\,\zeta(s)| \leqslant \{k(|t|+2)\}^{A_8\,n}, \tag{110}$$

for $-\frac{3}{2} \leqslant \sigma \leqslant \frac{11}{2}$.

A lower bound for $|\zeta(s)|$ is also required for $\sigma = 2$. This is given through (107) by

$$\frac{1}{|\zeta(s)|} \leqslant \prod_{\mathfrak{p}} \left(1 + \frac{1}{(N\mathfrak{p})^2}\right) \leqslant \zeta(2) \leqslant A_3^n,$$

from which it follows that

$$\frac{1}{|(s-1)\,\zeta(s)|} \leqslant A_3^n \tag{111}$$

for $\sigma = 2$.

The bounds just obtained are needed in order to obtain information about the vertical distribution of the complex zeros. Let $N(T)$ for any non-negative T denote the number of zeros ρ such that $0 \leqslant \gamma \leqslant T$. Then, defining $\nu(y)$ to be the number of zeros (not necessarily all 'complex') of $\xi(s) = (s-1)\,\zeta(s)$ in the circle with centre $2+iT$ and radius y, we have, by Jensen's theorem and then by (110) and (111),

$$\int_0^{\frac{7}{2}} \frac{\nu(y)}{y}\,dy = \frac{1}{2\pi}\int_0^{2\pi} \log|\xi(2+iT+\tfrac{7}{2}e^{i\theta})|\,d\theta - \log|\xi(2+iT)|$$

$$\leqslant A_9 n \log\{k(T+2)\}.$$

Next

$$\int_0^{\frac{7}{2}} \frac{\nu(y)\,dy}{y} \geqslant \int_3^{\frac{7}{2}} \frac{\nu(y)\,dy}{y} \geqslant \nu(3)\log\tfrac{7}{6},$$

and $\nu(3) \geqslant N(T+1)-N(T)$. We, therefore, conclude that

$$N(T+1)-N(T) = O\{n\log[k(T+2)]\}. \tag{112}$$

At this point it is necessary for the first time to assume the extended Riemann hypothesis, which we state explicitly thus.

HYPOTHESIS *The real part β of every complex zero $\rho = \beta+i\gamma$ of Dedekind's zeta function is equal to $\tfrac{1}{2}$ for every Kummer field of type $G = Q(\sqrt[k]{2}, \sqrt[k]{1})$.*

We estimate $\psi(x)$ through the explicit formula (iv), which is now written in the form

$$\psi(y) = y - a_0 - \sum_\rho \frac{y^\rho}{\rho} + O(n\log y) \tag{113}$$

for non-integral values of y that are not less than $\tfrac{3}{2}$. Assuming for the present that x is of the form $m+\tfrac{1}{2}$, where $m \geqslant 1$, let $x \leqslant y \leqslant x+\tfrac{1}{4}$. Then, since $\rho = \tfrac{1}{2}+i\gamma$ by hypothesis, we have

$$\left|\sum_{|\gamma|\leqslant x} \frac{y^\rho}{\rho}\right| \leqslant y^{\frac{1}{2}} \sum_{|\gamma|\leqslant x} \frac{1}{|\rho|} = O\left(x^{\frac{1}{2}} \sum_{0\leqslant\gamma\leqslant x} \frac{1}{\gamma+1}\right)$$

$$= O\left\{x^{\frac{1}{2}}\left(N(1)+\sum_{\mu=1}^m \frac{N(\mu+1)-N(\mu)}{\mu}\right)\right\}$$

$$= O\left\{nx^{\frac{1}{2}}\log kx.\left(1+\sum_{\mu=1}^m \frac{1}{\mu}\right)\right\}$$

$$= O\{nx^{\frac{1}{2}}\log kx\log x\}$$

from (112). Therefore, by this and (113),

$$\psi(x) = \psi(y) = y - a_0 - \sum_{|\gamma|>x} \frac{y^\rho}{\rho} + O(nx^{\frac{1}{2}}\log kx\log x),$$

which when integrated in the range $x \leqslant y \leqslant x + \frac{1}{4}$ gives

$$\psi(x) = x + \tfrac{1}{8} - a_0 + 4 \sum_{|\gamma| > x} \frac{x^{\rho+1} - (x + \tfrac{1}{4})^{\rho+1}}{\rho(\rho+1)} + O(nx^{\frac{1}{2}} \log kx \log x)$$

$$= x - a_0 + O(nx^{\frac{1}{2}} \log kx \log x) + O\left(x^{\frac{3}{2}} \sum_{\gamma > x} \frac{1}{\gamma^2} \right),$$

the term-by-term integration being justified by (iv). In this, by (112), we have

$$\sum_{\gamma > x} \frac{1}{\gamma^2} \leqslant \sum_{\mu \geqslant m} \frac{N(\mu+1) - N(\mu)}{\mu^2} = O\left(n \sum_{\mu \geqslant m} \frac{\log k\mu}{\mu^2} \right) = O\left(\frac{n \log kx}{x} \right).$$

Consequently

$$\psi(x) = x - a_0 + O(nx^{\frac{1}{2}} \log kx \log x).$$

Therefore, finally, avoiding the problem of determining the order of magnitude of a_0 directly, we infer that

$$\psi(x) = \psi(x) - \psi(\tfrac{3}{2}) = x + O(nx^{\frac{1}{2}} \log kx \log x), \qquad (114)$$

where the formula clearly remains valid for all $x \geqslant \frac{3}{2}$ after the successive removal of the restrictions that x should be of the form $x = m + \frac{1}{2}$ and that x should not be an integer.

The transition from $\psi(x)$ to $\Pi(x, k)$ is easily effected. Letting

$$\theta(x) = \sum_{N\mathfrak{p} \leqslant x} \log N\mathfrak{p},$$

we have

$$\theta(x) = \sum_{\alpha=1}^{\infty} \mu(\alpha)\, \psi(x^{1/\alpha})$$

in virtue of the obvious identity

$$\psi(x) = \sum_{\beta=1}^{\infty} \theta(x^{1/\beta}).$$

Thus $\psi(x)$ may be replaced by $\theta(x)$ in (114), whereupon it follows by partial summation that

$$\pi(x, k) = \mathrm{li}\, x + O(nx^{\frac{1}{2}} \log kx).$$

We then have from this and (105) the formula

$$P(x, k) = \frac{\mathrm{li}\, x}{k\phi(k)} + O(x^{\frac{1}{2}} \log kx), \qquad (115)$$

upon which the remainder of the investigation depends.

5   Estimation of $\mathrm{N}(x, \xi_1)$ and $\mathrm{M}(x, \xi_1, \xi_2)$

Taking $\mathrm{N}(x, \xi_1)$ first, we have from (94), (95), and (115) that

$$\mathrm{N}(x, \xi_1) = \sum_{l'} \mu(l') \left(\frac{\mathrm{li}\, x}{l'\phi(l')} + O(x^{\frac{1}{2}} \log x) \right)$$

$$= \mathrm{li}\, x \sum_{l'} \frac{\mu(l')}{l'\phi(l')} + O\left(\sum_{l \leqslant x^{\frac{1}{2}}} x^{\frac{1}{2}} \log x \right)$$

$$= \mathrm{li}\, x \sum_{l'} \frac{\mu(l')}{l'\phi(l')} + O\left(\frac{x}{\log^2 x} \right).$$

Therefore, since all square-free numbers k not exceeding ξ_1 are certainly of the type l', we have

$$\mathrm{N}(x, \xi_1) = \mathrm{li}\, x \sum_{k=1}^{\infty} \frac{\mu(k)}{k\phi(k)} + O\left(\mathrm{li}\, x \sum_{k > \xi_1} \frac{1}{k\phi(k)} \right) + O\left(\frac{x}{\log^2 x} \right)$$

$$= \frac{x}{\log x} \sum_{k=1}^{\infty} \frac{\mu(k)}{k\phi(k)} + O\left(\frac{x}{\xi_1 \log x} \right) + O\left(\frac{x}{\log^2 x} \right)$$

$$= \frac{Cx}{\log x} + O\left(\frac{x}{\log^2 x} \right), \tag{116}$$

where
$$C = \prod_{q} \left(1 - \frac{1}{q(q-1)} \right).$$

Passing on to $\mathrm{M}(x, \xi_1, \xi_2)$, we have at once from (96) and (115) that

$$\mathrm{M}(x, \xi_1, \xi_2) \leqslant \sum_{\xi_1 < q \leqslant \xi_2} \left(\frac{\mathrm{li}\, x}{q(q-1)} + O(x^{\frac{1}{2}} \log x) \right)$$

$$= O\left(\frac{x}{\log x} \sum_{q > \xi_1} \frac{1}{q^2} \right) + O\left(x^{\frac{1}{2}} \log x \sum_{q \leqslant \xi_2} 1 \right)$$

$$= O\left(\frac{x}{\xi_1 \log x} \right) + O\left(\frac{\{x^{\frac{1}{2}} \log x\} \xi_2}{\log \xi_2} \right) = O\left(\frac{x}{\log^2 x} \right), \tag{117}$$

which is the estimate required.

6   The theorem

The theorem is obtained by combining the results of equations (93), (116), and (117), part (*b*) being a corollary of part (*a*).

THEOREM 2   *If it is assumed that the extended Riemann hypo-thesis hold for the Dedekind zeta function over Kummer fields of the type $Q(\sqrt[k]{2}, \sqrt[k]{1})$, where k is square-free, then we have:*

(a) *let $N_2(x)$ be the number of primes p not exceeding x for which 2 is a primitive root, modulo p; then*

$$N_2(x) = \frac{Cx}{\log x} + O\left(\frac{x \log \log x}{\log^2 x}\right),$$

where
$$C = \prod_q \left(1 - \frac{1}{q(q-1)}\right);$$

(b) *there are infinitely many primes p for which 2 is a primitive root, modulo p.*

7   Relaxation of hypotheses

In our quest for an unconditional proof of the Artin conjectures it is natural to begin by considering what can be achieved on the basis of the usual form of the prime ideal theorem. When inter-preted through the notation of § 3 this theorem gives

$$\pi(x, k) = \operatorname{li} x + O(x e^{-c_k \sqrt{\log x}}),$$

where the constant implied by the O-notation may depend on k. This in turn leads through (105) to the formula

$$P(x, k) = \frac{\operatorname{li} x}{k \phi(k)} + O(x e^{-c_k \sqrt{\log x}}), \tag{118}$$

with a similar understanding with regard to the O-notation.

An unconditional proof readily follows for the upper bound that is inherent in the asymptotic formula given by Theorem 2. Indeed, using (94) and (118) in conjunction with a method that is little more than an amplification of some remarks made in Chapter 1, § 1, about the simple asymptotic sieve, we easily find that

$$N_2(x) \leqslant N(x, \xi_1) = \sum_{l'} \mu(l') P(x, l') \leqslant \frac{(1+\eta) Cx}{\log x},$$

for $x > x_0(\eta)$, provided ξ_1 is now chosen to be a function of x that tends sufficiently slowly to infinity as $x \to \infty$. Although only recently formally enunciated and proved by Vinogradov [80], this result has been widely known for many years (see, for example, Heilbronn in [55]).

Returning to the deeper question of the asymptotic formula itself, we also see immediately that $M(x; \xi_1, \xi_2)$ is now the only term in the right-hand side of (90) for which a satisfactory unconditional estimate has not yet been obtained for an appropriate value of ξ_1. Since the previous estimation of this term was based through (96) on sums $P(x, q)$ with prime values q, it then follows that we need only assume the Riemann hypothesis in respect of zeta functions over fields of the type $G = Q(\sqrt[q]{2}, \sqrt[q]{1})$, the conclusions of Theorem 2 still remaining valid so long as the error term in the asymptotic formula is appropriately modified.

The conditional aspects of the result may be further reduced by relating the assumptions to the fields $F = Q(\sqrt[q]{2})$ rather than to their relative extensions G. To achieve this further improvement we reconsider the criteria given in § 3 for determining whether 2 be a qth power residue, $\bmod p$. The modified criteria are then applied to the problem through the theory of prime ideals in the fields F, it being convenient to preface the argument by the introduction of the following notation: $\Pi(x, q)$ is the number of prime ideals in F whose norms do not exceed x; $\Pi^{(1)}(x, q)$ is the contribution to $\Pi(x, q)$ due to linear prime ideals not dividing $2q$; and $\Pi^{(2)}(x, q)$ is the complementary contribution (cf. the definitions leading to (105)).

We mentioned in § 3 that the simultaneous conditions that $v^q \equiv 2, \bmod p$, be soluble and that $p \equiv 1, \bmod q$, were equivalent to the condition that $v^q \equiv 2, \bmod p$, have q incongruent roots, $\bmod p$. To this we now add the remark that $v^q \equiv 2, \bmod p$, always has exactly one root, $\bmod p$, when $p \not\equiv 1, \bmod q$, and $p \neq q$ (the condition $p \neq 2$ remaining implicit throughout as before). From these statements it then follows by Dedekind's principle that

$$\Pi^{(1)}(x, q) = qP(x, q) + \sum_{\substack{p \leqslant x;\ p \neq 2,\ q \\ p \not\equiv 1,\ \bmod q}} 1,$$

while trivially

$$\Pi^{(2)}(x, q) \leqslant 2q + q\pi(x^{\frac{1}{2}}) = O(qx^{\frac{1}{2}}).$$

Therefore
$$qP(x, q) = \Pi(x, q) - \pi(x) + \pi(x; 1, q) + O(qx^{\frac{1}{2}}), \qquad (119)$$

whose constituent terms on the right-hand side are estimated as follows.

First, assuming the Riemann hypothesis for the zeta functions over the fields F, we have

$$\Pi(x, q) = \operatorname{li} x + (qx^{\frac{1}{2}} \log qx) \qquad (120)$$

by a simple adaptation of the method of §4. Secondly

$$\pi(x) = \operatorname{li} x + O\left(\frac{x}{\log^2 x}\right) \qquad (121)$$

by the prime number theorem.† Lastly, for $q \leqslant \xi_2$,

$$\pi(x; 1, q) = O\left(\frac{x}{q \log x}\right) \qquad (122)$$

by Lemma 1.

We infer from (119), (120), (121), and (122) that

$$P(x, q) = O(x^{\frac{1}{2}} \log x) + O\left(\frac{x}{q \log^2 x}\right) + O\left(\frac{x}{q^2 \log x}\right)$$

for $q \leqslant \xi_2$, and hence, finally, that

$$
\begin{aligned}
M(x; \xi_1, \xi_2) &\leqslant \sum_{\xi_1 < q \leqslant \xi_2} P(x, q) \\
&= O\{x^{\frac{1}{2}} \log x \cdot \pi(\xi_2)\} + O\left(\frac{x}{\log^2 x} \sum_{q \leqslant \xi_2} \frac{1}{q}\right) + O\left(\frac{x}{\log x} \sum_{q > \xi_1} \frac{1}{q^2}\right) \\
&= O\left(\frac{x \log \log x}{\log^2 x}\right) + O\left(\frac{x}{\xi_1 \log x}\right) = o\left(\frac{x}{\log x}\right)
\end{aligned}
$$

as $x \to \infty$. In view of earlier remarks this shews that the Artin asymptotic formula holds if the Riemann hypothesis is true for the zeta functions over the fields $F = Q(\sqrt[q]{2})$.

This simple method can be directly compared with the Vinogradov treatment [80] to which we adverted in §1. If the

---

† The remainder term can be substantially improved because the Riemann hypothesis for the zeta function over any field F implies the Riemann hypothesis for $\zeta(s)$ (cf. remarks in a later paragraph). The formula given here, however, suffices for our purposes, and we wish to avoid appealing to the theory that makes the improvement possible.

zeta function over any number field M is denoted by $\zeta_M(s)$, then, as Vinogradov points out, we have the identities

$$\zeta_G(s) = \zeta_Z(s) \, L^{q-1}(s, \chi_q),$$

$$\zeta_F(s) = \zeta(s) \, L(s, \chi_q),$$

and
$$\zeta_Z(s) = \zeta(s) \prod_{\chi \neq \chi_0} L(s, \chi),$$

where $L(s, \chi)$ is a Dirichlet's L function to modulus q and $L(s, \chi_q)$ is an Artin's L function that in this case is an integral function. Thus the zeros of $\zeta_G(s)$ are either zeros of $\zeta(s)$, or $L(s, \chi)$, or $L(s, \chi_q)$. The substance of the latter part of Vinogradov's method is then to the effect that we need only hypothesize on the distribution of the zeros of the functions $L(s, \chi_q)$ since the contribution due to the zeros of the Dirichlet L functions can be handled by the zero density theorems that led to Bombieri's theorem (cf. comments immediately after Lemma 1 on p. 10). Similarly our method shews that it is enough to work in terms of $\zeta_F(s)$ and $\pi(x; 1, q)$, where $\zeta_F(s)$ corresponds to all intents and purposes to $L(s, \chi_q)$ and where $\pi(x; 1, q)$ corresponds to the Dirichlet L functions $L(s, \chi)$. It should be observed, however, that our method depends on the Brun–Titchmarsh theorem rather than on the deeper Bombieri theorem that is used by Vinogradov.

There are a multitude of ways in which the hypotheses used can be further weakened. One such way, as Vinogradov suggests, is to substitute zero density hypotheses for the Riemann hypothesis. Although it would not be profitable to expand on this matter further, it may perhaps be of interest to end this chapter with the observation that the second part of Theorem 2 is still true if no zero of the zeta functions over F has real part exceeding $1 - \frac{1}{2}e^{-1} - \delta \; (> \frac{4}{5})$.

4. Power-free values of polynomials – a joint application of the simple asymptotic sieve and the large sieve

1 Introduction

The simple asymptotic sieve is particularly appropriate for the treatment of questions involving the representation of lth-power-free numbers (abbreviated in the sequel to 'l-free numbers') by polynomials of degree $r \leqslant l$. In the somewhat voluminous literature on this subject the central theorem is due to Ricci [64], who in 1933 used a form of Brun's method to prove that

$$N(x) = N(x, f, l) \sim A(f, l) x \quad (A > 0),$$

where $N(x, f, l)$ is the number of integers n not exceeding x for which the primitive irreducible polynomial $f(n)$ of degree $r \leqslant l$ is l-free. Other writers tended to consider special cases of particular interest, their methods being similar in scope to that of the simple asymptotic sieve.

When $l < r$ a straightforward application of the asymptotic sieve no longer leads to a solution, since the set of sieving prime-powers becomes too dense for the remainder terms in the process to be satisfactorily assessed. The first progress in this case was made as late as 1953 when Erdös [15] proved by an ingenious method that an irreducible integral polynomial of degree $r \geqslant 3$ represents $(r-1)$-free numbers infinitely often, provided that the obvious necessary condition is given that $f(n)$ have no fixed $(r-1)$th power divisors other than 1 (the corresponding condition when $r \geqslant l$ was not stated earlier because in that case it always holds when $f(n)$ is primitive). His method did not give a means for determining an asymptotic formula for $N(x, f, r-1)$, nor did it shew that the integers n for which $f(n)$ is $(r-1)$-free had positive upper density. In 1966, however, after having first considered a special case through the study of a relevant

Diophantine equation [38], the author [35] proved the asymptotic formula

$$N(x, f, r-1) = A(f, r-1)x + O(x \log^{-A_1/\log \log \log x} x), \quad (123)$$

by an alternative method. His procedure involved the combination of the simple asymptotic method with an idea having some affinity with the large sieve.

We have throughout limited our description to the situation of greatest interest where the polynomial is irreducible, since the reducible case is easier and is largely dependent on the irreducible one. Indeed when the polynomial is reducible it is easy to make further progress with these problems by utilizing the fact that the irreducible factors of the polynomial have degree less than r. In contrast the case $l < r-1$ seems still to present unsurmountable difficulties for irreducible polynomials even though there is every expectation that the corresponding results are true. Another problem of the same type involves the representation of l-free numbers by $f(p)$, where p is a prime. Here the problem is still manageable for $l > r$ and just manageable for $l = r$ (for the latter case, see, for example, Uchiyama [78]) but so far no published result has been given for $l = r-1$. Yet Erdös has conjectured that $f(p)$ infinitely often represents l-free numbers in all situations where the obvious necessary conditions obtain [15], [17].

Similar problems are presented by the conjugate question of the representation of large numbers M as the sum of an l-free number and a value of the polynomial $f(n)$, the most interesting cases being those in which $f(n)$ is taken to be n^r; here it is no longer appropriate to insist that $f(n)$ be irreducible since it is now the representation of power-free numbers by $M - f(n)$ that is under consideration. Whether or not the argument of $f(n)$ be restricted to be a prime, counterparts of all the results hitherto mentioned can be derived in this context when $l \geqslant r$. On the other hand no such results have hitherto been established for the case $l = r-1$, since arithmetical difficulties associated with the largeness of the integers M preclude the direct adaptation of both Erdös's method and the author's to meet this situation.†

† Erdös almost certainly overlooked these arithmetical complications when he asserted in [15] that his method was applicable to this case. We should

In this chapter we illustrate the use of the simple asymptotic sieve in combination with the large sieve by proving (123) in the improved form

$$N(x, f, r-1) = A(f, r-1)x + O(x \log^{\{2/(r+1)\}-1} x) \quad (r \geqslant 3).$$

We then discuss briefly various elaborations of the method that have a relevance to the unsolved problems mentioned above. This will lead first to the conclusion that there are indeed some irreducible polynomials of degree r for which $f(p)$ is infinitely often $(r-1)$-free. It will also enable us to deduce the solutions of some outstanding conjugate problems including that of the representation of large numbers as the sum of an $(r-1)$-free number and an rth power. Finally we shall mention how these ideas lead also to results that are of interest in connection with Waring's problem for cubes.

2   Application of the simple asymptotic sieve

We begin by noting that the polynomial $f(n)$ never vanishes in virtue of its irreducibility; furthermore we may suppose without loss of generality that the leading coefficient of $f(n)$ is positive so that $f(n)$ is positive for all but at most a finite number of values of n.

We require a notation for sums that count the number of positive integers n not exceeding x for which $f(n)$ has appropriate properties. First $N'(x)$ appertains to the property that $f(n)$ be not divisible by the $(r-1)$th power of any prime number not exceeding ξ_1, where $\xi_1 = \frac{1}{6}\log x$; next $N''(x)$ appertains to the property that $f(n)$ be divisible by the $(r-1)$th power of at least one prime number exceeding ξ_1; while, finally, $N_l(x)$ appertains to the property that $f(n)$ be divisible by l.

We use the basic relation

$$N(x) = N'(x) + O\{N''(x)\} \tag{124}$$

that was established in Chapter 1, §1.

also take this opportunity to mention that a proof of an asymptotic formula was attempted by Moĭšezon and Subhankulov for the case in which $f(n) = n^r$ without restriction on r and l [56]; their demonstration, however, was erroneous and depended on a method that is not likely to make any substantial contribution to problems of this nature.

Taking $N'(x)$ first, let l' indicate, generally, either 1 or square-free numbers composed entirely of prime factors not exceeding ξ_1. Then, by following the method used to derive (38), we have

$$N'(x) = \sum_{l'} \mu(l') N_{l'^{(r-1)}}(x)$$

$$= \sum_{l'} \mu(l') \rho(l'^{(r-1)}) \left(\frac{x}{l'^{(r-1)}} + O(1) \right),$$

where $\rho(l)$ denotes the number of (incongruent) roots of the congruence $f(\nu) \equiv 0, \bmod l$. Since $l' \leqslant x^{\frac{1}{3}}$ as in (95), and since

$$\rho(l'^{(r-1)}) = O\{d_r(l')\}$$

by the elementary theory of congruences, we conclude that

$$N'(x) = x \sum_{l'} \frac{\mu(l') \rho(l'^{(r-1)})}{l'^{(r-1)}} + O(\sum_{l \leqslant x^{\frac{1}{3}}} d_r(l))$$

$$= x \prod_{p \leqslant \xi_1} \left(1 - \frac{\rho(p^{r-1})}{p^{r-1}} \right) + O(x^{\frac{1}{3}} \log^{(r-1)} x)$$

$$= x \prod_p \left(1 - \frac{\rho(p^{r-1})}{p^{r-1}} \right) + O\left(\frac{x}{\xi_1^{r-2} \log \xi_1} \right) + O(x^{\frac{1}{3}} \log^{(r-1)} x)$$

$$= x \prod_p \left(1 - \frac{\rho(p^{r-1})}{p^{r-1}} \right) + O\left(\frac{x}{\log x} \right). \tag{125}$$

Next, if $\xi_2 = x \log^{2/(r+1)} x$, then

$$N''(x) \leqslant \sum_{p > \xi_1} N_{p^{r-1}}(x) = \sum_{\xi_1 < p \leqslant \xi_2} + \sum_{p > \xi_2} = P_1(x) + P_2(x), \quad \text{say.} \tag{126}$$

The sum $P_1(x)$ can be estimated immediately. We have

$$P_1(x) = \sum_{\xi_1 < p \leqslant \xi_2} \rho(p^{r-1}) \left(\frac{x}{p^{r-1}} + O(1) \right) = O\left(x \sum_{p > \xi_1} \frac{1}{p^{r-1}} \right) + O(\sum_{p \leqslant \xi_2} 1)$$

$$= O\left(\frac{x}{\xi_1^{r-2} \log \xi_1} \right) + O\{\pi(\xi_2)\}$$

$$= O\left(\frac{x}{\log x} \right) + O\left(\frac{x}{\log^{1-2/(r+1)} x} \right)$$

$$= O(x \log^{\{2/(r+1)\}-1} x). \tag{127}$$

The estimation of $P_2(x)$ is longer, constituting in fact the main difficulty to be overcome. We end this section by transforming this sum into a double sum in preparation for the investigation in the next section.

Evidently
$$P_2(x) = \sum_{\substack{f(n)=\mu p^{r-1} \\ n \leqslant x;\, p > \xi_2}} 1,$$

the conditions of summation and the irreducibility of $f(n)$ implying that

$$-1 < -\frac{A_2}{\xi_2^{r-1}} < \mu < \frac{A_3 x^r}{\xi_2^{r-1}} = A_3 x \log^{-2(r-1)/(r+1)} x; \quad \mu \neq 0.$$

Therefore, setting $\xi_3 = A_3 x \log^{-2(r-1)/(r+1)} x$, we obtain

$$P_2(x) = \sum_{m < \xi_3} \sum_{\substack{f(n)=mp^{r-1} \\ n \leqslant x;\, p > \xi_2}} 1 = \sum_{m < \xi_3} \Upsilon(m), \quad \text{say.} \tag{128}$$

3 Application of the large sieve to $\Upsilon(m)$

The condition $m < \xi_3$ given in (128) will remain implicit throughout this section.

Let q be any prime such that $A_4 < q \leqslant x$, $q \nmid m$, and $q \equiv 1$, mod $(r-1)$. Then the upper bound for $\Upsilon(m)$ is derived through the consideration that p^{r-1} in the conditions of summation defining $\Upsilon(m)$ is an $(r-1)$th power residue, mod q; furthermore this power residue is relatively prime to q in virtue of $p > q$.

First n must be confined to $\rho(m)$ residue classes, mod m, in the sum defining $\Upsilon(m)$. Next, considering those values of n that belong to one such given residue class, mod m, we have the additional condition that

$$f(n) \equiv ma_q, \bmod q, \tag{129}$$

for some (proper) $(r-1)$th power residue, a_q, mod q. This shews that
$$n \equiv \nu, \bmod q,$$

where ν appears in a solution of the congruences

$$f(\nu) \equiv m\mu^{r-1}, \bmod q; \quad \mu \not\equiv 0, \bmod q. \tag{130}$$

Let (130) have $S(m, q)$ incongruent solutions, mod q, in μ, ν. Then, since a_q has $r-1$ representations, mod q, in the form μ^{r-1}, we see through (129) that n is confined to

$$\frac{1}{r-1} S(m, q)$$

residue classes, mod q. Moreover, since $q \nmid m$ and $f(n)$ is irreducible, a theorem of Weil [82] shews that

$$S(m, q) = q + O(q^{\frac{1}{2}}) > \tfrac{1}{2}q \tag{131}$$

for a suitable value of A_4.

The position has been prepared for the use of Gallagher's form of the large sieve. Still restricting our attention to the contribution to $\Upsilon(m)$ due to those n belonging to a given residue class, mod m, we obtain an upper bound for this contribution by defining $v(l)$ in Lemma 3 on p. 19 for prime-powers l by

$$
v(l) = \begin{cases}
1, & \text{if } l \mid m, \\[2mm]
\dfrac{1}{r-1}S(m, q), & \text{if } l \text{ is a prime of type } q, \\[2mm]
l, & \text{if } l \text{ is any other prime-power,}
\end{cases}
$$

and by defining T to be the set of prime-power divisors of m together with the primes p such that $p \nmid m$ and $A_4 < p \leqslant X$, where in due course X is to be assigned an appropriate value less than x. Taking the denominator in the expression giving the upper bound in Lemma 3, we consider

$$
\sum_{l \in T} \frac{\Lambda(l)}{v(l)} - \log x = \sum_{l \mid m} \Lambda(l) + (r-1) \sum_{q \leqslant X} \frac{\log q}{S(m, q)}
$$
$$
+ \sum_{\substack{p \not\equiv 1,\, \mathrm{mod}\,(r-1) \\ A_4 < p \leqslant X;\, p \nmid m}} \frac{\log p}{p} - \log x,
$$

which by (131) is equal to

$$
(r-1) \sum_{\substack{p \equiv 1,\, \mathrm{mod}\,(r-1) \\ p \leqslant X}} \frac{\log p}{p} + \sum_{\substack{p \not\equiv 1,\, \mathrm{mod}\,(r-1) \\ p \leqslant X}} \frac{\log p}{p}
$$
$$
- \log \frac{x}{m} - (r-1) \sum_{\substack{p \equiv 1,\, \mathrm{mod}\,(r-1) \\ p \leqslant X;\, p \mid m}} \frac{\log p}{p}
$$
$$
- \sum_{\substack{p \not\equiv 1,\, \mathrm{mod}\,(r-1) \\ p \leqslant X;\, p \mid m}} \frac{\log p}{p} + O\!\left(\sum_p \frac{\log p}{p^{\frac{3}{2}}}\right) + O(1)
$$
$$
\geqslant \left(1 + \frac{r-2}{\phi(r-1)}\right) \log X - \log \frac{x}{m} - (r-1) \sum_{p \mid m} \frac{\log p}{p} + O(1)
$$
$$
> 2 \log X - \log \frac{x}{m} - A_5 - (r-1) \sum_{p \mid m} \frac{\log p}{p}.
$$

Therefore the denominator is certainly greater than 1 for

$$X = A_6 \left(\frac{x}{m}\right)^{\frac{1}{2}} \exp\left(\frac{(r-1)}{2} \sum_{p\mid m} \frac{\log p}{p}\right).$$

With this value of X the corresponding numerator in Lemma 3 is then

$$\sum_{l\mid m} \Lambda(l) + \sum_{\substack{A_4 < p \leqslant X \\ p\nmid m}} \log p - \log x \leqslant \sum_{p\leqslant X} \log p - \log\frac{x}{m} = O(X)$$

$$= O\left\{\left(\frac{x}{m}\right)^{\frac{1}{2}} \prod_{p\mid m}\left(1 + \frac{A_7 \log p}{p}\right)\right\}$$

$$= O\left\{\left(\frac{x}{m}\right)^{\frac{1}{2}} \prod_{p\mid m}\left(1 + \frac{1}{p^{\frac{1}{2}}}\right)\right\} = O\left\{\left(\frac{x}{m}\right)^{\frac{1}{2}} \sigma_{-\frac{1}{2}}(m)\right\},$$

and this is an upper bound for the contribution to $\Upsilon(m)$ due to the values of n that lie in a given residue class, mod m. Hence we deduce the required estimate

$$\Upsilon(m) = O\left\{\left(\frac{x}{m}\right)^{\frac{1}{2}} \sigma_{-\frac{1}{2}}(m)\rho(m)\right\}. \tag{132}$$

4 The proof completed

It remains to estimate $P_2(x)$ by means of (128) and (132). To this end we require for $u \geqslant 1$ the inequality

$$\sum_{m\leqslant u} \rho(m) = O(u), \tag{133}$$

which, as Erdös [13] and others have shewn by the theory of ideals, is a consequence of the irreducibility of $f(n)$. We also require the inequality

$$\rho(lm) = O\{\rho(l)\rho(m)\} \tag{134}$$

that follows from the elementary theory of congruences (see Nagell [62], Chapter 3).

We deduce from (133) and (134) that

$$\sum_{m\leqslant y} \sigma_{-\frac{1}{2}}(m)\rho(m) = \sum_{\lambda\mu\leqslant y} \frac{\rho(\lambda\mu)}{\lambda^{\frac{1}{2}}} = O\left(\sum_{\lambda\mu\leqslant y} \frac{\rho(\lambda)\rho(\mu)}{\lambda^{\frac{1}{2}}}\right)$$

$$= O\left(\sum_{\lambda\leqslant y} \frac{\rho(\lambda)}{\lambda^{\frac{1}{2}}} \sum_{\mu\leqslant y/\lambda} \rho(\mu)\right)$$

$$= O\left(y\sum_{\lambda\leqslant y} \frac{\rho(\lambda)}{\lambda^{\frac{3}{2}}}\right) = O(y). \tag{135}$$

Therefore, by (128) and (132) and then by (135) and partial summation, we have

$$P_2(x) = O\left(x^{\frac{1}{2}} \sum_{m \leqslant \xi_2} \frac{\sigma_{-\frac{1}{2}}(m)\rho(m)}{m^{\frac{1}{2}}}\right) = O(x^{\frac{1}{2}}\xi_3^{\frac{1}{2}}) = O(x\log^{\{2/(r+1)\}-1}x).$$

$$(136)$$

Returning to $N''(x)$ we infer at once from (126), (127), and (136) that

$$N''(x) = O(x\log^{\{2/(r+1)\}-1}x).$$

Taken with (124) and (125) this estimate implies the first part of the following theorem, the second part being an immediate corollary.

THEOREM 3   *Let $f(n)$ be an irreducible integral polynomial of degree $r \geqslant 3$. Then, if $N(x)$ is the number of positive integers n not exceeding x with the property that $f(n)$ is $(r-1)$-free, we have as $x \to \infty$*

$$N(x) = x\prod_p\left(1 - \frac{\rho(p^{r-1})}{p^{r-1}}\right) + O(x\log^{\{2/(r+1)\}-1}x).$$

If $f(n)$ has no fixed $(r-1)$th power divisors other than 1, then there are infinitely many integers n for which $f(n)$ is $(r-1)$-free.

5   The problem when the polynomial has prime argument

As before we consider the $(r-1)$-free numbers represented by the polynomial, but we restrict the argument of the polynomial to prime values. We define the analogues of $N'(x)$, $N''(x)$, $P_1(x)$, $P_2(x)$ in the obvious way save that ξ_2 must now be taken so that $\xi_2 = o(x)$. Then it is comparatively easy to prove that the analogue of $N'(x)$ is asymptotic to $Bx/\log x$ $(B > 0)$ using the prime number theorem for arithmetic progressions, while an elementary argument is sufficient to shew that the analogue of $P_1(x)$ is $o(x/\log x)$. On the other hand there does not seem to be any way in which we can estimate the analogue of $P_2(x)$ by taking advantage of the fact that the argument in the polynomial is a prime. Consequently, the problem turns out to be more or less equivalent to that of shewing that it is possible to improve the error term in Theorem 3 to the form $o(x/\log x)$, i.e. it is necessary to shew that $P_2(x)$ itself (the only consequential change in its definition being that $\xi_3/x \to \infty$) is $o(x/\log x)$.

For the estimation of $\Upsilon(m)$ in $P_2(x)$ Montgomery's form of the large sieve method would serve as well as Gallagher's. However, when $m > x$ no method can be expected to evaluate $\Upsilon(m)$ usefully for a given m, since then the range of n is small compared with m and the number of admissible values of n is at most $\rho(m)$. Nevertheless, since it was stated in Chapter 2, §9, that the roots of $f(\nu) \equiv 0$, mod m, have some uniformity in their distribution for variable m [33], it is not unreasonable to expect that a large sieving process might be able to estimate sums of the type $\Sigma \Upsilon(m)$ successfully. Here, in accordance with our comments in §4 of Chapter 1, the structure of the Gallagher and Montgomery forms of the large sieve does not readily permit different values of m to be taken together. On the other hand the Selberg method leads here to upper bounds of the form

$$\sum_d \tau_d \sum_{\substack{b_d \\ (m,\,d)=1}} \left(\frac{x\rho(m)\,\gamma(d, b_d, m)}{md} + R(x, d, b_d, m) \right) = \Sigma_1 + \Sigma_2, \quad \text{say,}$$

where the significance of the notation is as follows: (i) d belongs to a suitable class of square-free numbers composed of prime factors q that are congruent to 1, mod $(r-1)$; (ii) b_d runs through all reduced residues, mod d, that are congruent to $(r-1)$th power non-residues, modulis all prime divisors q of d; (iii) $R(x, d, b_d, m)$ is a remainder term; (iv) τ_d is given in terms of the appropriate Selberg lambda coefficients; and (v) $\gamma(d, b_d, m)$ is the number of roots of the congruence

$$f(n) \equiv mb_d, \bmod d.$$

The treatment of Σ_1 does not involve the introduction of any fresh principles. In fact, the dominant part of the inner sum in Σ_1 may be shewn to equal

$$x \left(\frac{r-2}{r-1} \right)^{\omega(d)} \sum_{(m,\,d)=1} \frac{\rho(m)}{m}$$

by a modification of the reasoning previously used in connection with (129), (130), and (131). Summation over d then yields a satisfactory upper bound for Σ_1 provided that the admissible range of d is sufficiently wide.

An appropriate estimate for Σ_2 is now needed in a situation where the ranges of summation in m and d are compatible with the initial conditions and with an adequate assessment of Σ_1. To consider this sum in such circumstances, we utilize the fact that the remainder term $R(x, d, b_d, m)$ can be expressed as a Fourier series by the method of Chapter 2, and that summation over b_d then leads to an expression involving exponential sums with denominator δm, where $\delta | d$. Each such exponential sum can be written as a product of two exponential sums, one with denominator δ and the other with denominator m. The non-trivial evaluation of the first exponential sum involves a generalized Gaussian sum, and gives rise in itself to a result equivalent to that obtainable by the Montgomery or Gallagher sieve (cf. remarks made on the large sieve in Chapter 1, §4). But the second exponential sum is of the general form

$$\sum_{f(\nu) \equiv 0, \bmod m} e^{2\pi i h \nu / m},$$

and an additional saving is achieved by using the non-trivial estimates that are provided for

$$\sum_m \Big| \sum_{\substack{f(\nu) \equiv 0, \bmod m \\ 0 < \nu \leqslant m}} e^{2\pi i h \nu / m} \Big|$$

by [33].

The use of the Selberg method here always enables some improvement in Theorem 3 to be made. Moreover, if $f(n)$ is a *normal* polynomial, then it leads to the conclusion that $f(p)$ is infinitely often $(r-1)$-free provided $r \geqslant 49$; if, furthermore, $f(n)$ is Abelian, then the conclusion holds provided $r \geqslant 31$.

As foreshadowed in §9, Chapter 2, further progress of a conditional nature can be achieved by combining these ideas with other methods for treating the roots of polynomial congruences. Thus, by an elaborate refinement of the analysis given at the end of that section, it can be demonstrated that

$$p^3 - 2$$

is infinitely often square-free so long as certain incomplete sums of Kloosterman type are sufficiently small. A similar procedure would also yield a conditional variant of Theorem 3 in which the remainder term would be of the form $O(n^{1-\delta})$ when $f(n) = n^3 - 2$, extensions to cover the cases of other polynomials being in principle possible.

When $f(n)$ takes certain special forms it is sometimes easier to obtain results by using the theory of Diophantine equations to consider the number of solutions of $ml^{r-1} = f(n)$ for a given value of m. For instance, it can be shewn by such means that

$$p^{16} + 1$$

is infinitely often 15th-power-free.

6　The conjugate problem

Since the representation of large numbers M as the sum of a positive $(r-1)$-free number and a positive value of $f(n)$ is equivalent to the representation of positive $(r-1)$-free numbers by the polynomial $g_M(n) = M - f(n)$, it is natural to consider the relevance of the methods of §§ 2 and 3 to the conjugate problem. Yet a new feature is presented in that the admissible values of n are now limited to those for which $g_M(n)$ is positive, the consequence being that x must be replaced by a function of M that is asymptotic to a positive multiple of $M^{1/r}$ as $M \to \infty$. After all minor adjustments consequent upon this change have been made, one can easily verify that the analysis of § 2 can be adapted to meet the new situation. In contrast the treatment of § 3 does not lend itself to any obvious modification because it is not yet known whether inequalities of type (133) are true when u is a fairly small function of the discriminant of the polynomial that corresponds to $\rho(m)$. Indeed, although apposite conditional results similar to (133) can be established by the methods of Chapter 3 on assuming the Riemann hypothesis for zeta functions over suitable algebraic number fields, the most that is known both unconditionally and uniformly is the inadequate inequality derived by replacing $\rho(m)$ by $d_r(m)$. It is therefore necessary to have recourse to deeper methods.

In some of the more interesting special cases the above mentioned difficulties can be overcome by combining the methods of §5 with other ideas. In particular, we can successfully treat the most important case where $f(n) = n^r$, and are led to the following

THEOREM 4 *All sufficiently large numbers are the sum of a positive $(r-1)$-free number and a positive rth power.*

This theorem should be compared with Estermann's result about numbers representable as the sum of a square-free number and a square [18].

7 Another application of the large sieve

We have been using the large sieve to estimate expressions defined in terms of powers by making use of the obvious property that lth powers are not lth power non-residues, modulis suitable primes. This idea has other applications. For instance, it can be shewn that the number of representations of a large number n as the sum of four positive cubes is $O(n^{\frac{2}{3}-\frac{1}{18}+\epsilon})$, this being the first improvement over the estimate $O(n^{\frac{2}{3}})$ that has been discovered ($O(n^{\frac{2}{3}+\epsilon})$ is trivial, while the ϵ in the exponent can be removed by appealing to the author's asymptotic formula for $\sum_{n\leqslant x} r_*^2(n)$, where $r_*(n)$ is the number of representations of n as the sum of two positive cubes [32]). Here we write the Diophantine equation in the form

$$r(r^2 + 3s^2) = 4n - 4X^3 - 4Y^3,$$

and then replace the square s^2 by a sequence generated by an appropriate large sieving process. The Selberg method is then applied much as in the previous section, the use of exponential sums being both necessary and complicated.

An application of these ideas also occurs in the theory of binary cubic forms. This leads to the theorem that, if $f(x,y)$ is a binary cubic form with integral coefficients and with determinant not equal to $-3\Delta^2$, then almost all numbers representable by $f(x,y)$ are representable in (essentially) only one way (Hooley [37a]).

5. *A problem of Hardy and Littlewood – an application of the enveloping sieve (and the upper bound sieve)*

1 Introduction

In a well-known paper Hardy and Littlewood [28] stated an asymptotic formula, suggested by a merely formal application of their circle method, for the number of representations of a number n as the sum of two squares and a prime; the truth of this formula would imply that every sufficiently large number was the sum of two squares and a prime. In a later paper they furthermore suggested that on the extended Riemann hypothesis (this, which here appertains to Dirichlet's L-functions, we hereafter refer to as Hypothesis R) it should be possible to prove that *almost all* numbers could be so represented. The latter proof was effected by Miss Stanley [75] through the use of the circle method, as were proofs (also on Hypothesis R) of asymptotic formulae for the number of representations of a number as sums of greater numbers of squares and primes, the dependence of her results on the unproved hypothesis being afterwards removed by Estermann [19] and others.

No progress, however, was made with the original problem of Hardy and Littlewood until the author shewed in 1957 [31] that their asymptotic formula was true on Hypothesis R. His method was founded on the fact that the number of representations of n in the required form is equal to the sum

$$\sum_{p\leqslant n} r(n-p),$$

where $r(\nu)$ denotes the number of representations of ν as the sum of two squares. Since $r(\nu)$ may be expressed as a sum over the divisors of ν, he noted that this sum is related in character to the sum

$$\sum_{0<p+a\leqslant x} d(p+a),$$

for which an asymptotic formula had been obtained by Titchmarsh [76] on Hypothesis R. He also remarked that the two sums were similar in that each could be expressed as a combination of terms of the type $\pi(m; b, k)$, where k belongs to a certain range that depends on the parameter m, in each case the strength of Hypothesis R being sufficient to estimate $\pi(m; b, k)$ over nearly all the required range of k. In the Titchmarsh problem the contribution due to the remainder of the range of k had been easily assessed by means of the upper bound given for $\pi(m; b, k)$ by Brun's method (whence the naming of Lemma 1 as the Brun–Titchmarsh theorem), the order of magnitude of this contribution being negligible in comparison with the main term in the resulting asymptotic formula. Such a straightforward procedure was, however, ineffective for the Hardy–Littlewood problem because, in order to obtain an estimate of a sufficiently small size, it was essential to take into account the changes of sign due to the presence of the quadratic character in the expression for $r(\nu)$. To meet the latter requirement the author devised an elaborate method in which asymptotic formulae, and not merely upper bounds, were needed for sums defined in terms of an invariant sieving process. Intrinsically the procedure consisted of replacing the primes in the problem at the appropriate stage by a fixed set of numbers containing them – whence the term *enveloping sieve* that was later used by Linnik to describe the method. The author also pointed out that the sum

$$\sum_{0 < p+a \leqslant x} r(p+a)$$

could be treated conditionally in like manner. The corresponding asymptotic formula was stated, and this shewed that there would be infinitely many primes of the form $u^2 + v^2 + a$ for any fixed non-zero integer a. Finally in commenting on the conditional nature of his proofs he observed that Hypothesis R was only needed in order to make available satisfactory formulae for $\pi(m; b, k)$.

Three years later Linnik [52] established unconditionally the truth of the Hardy–Littlewood formula. Although based in part on [31] and in particular on the enveloping sieve, Linnik's paper was noteworthy in its use of the powerful *dispersion method*, which Linnik has subsequently described in his book *The disper-sion method in binary additive problems* [53]. Also by a simpler application of this method Linnik found an unconditional solu-tion of Titchmarsh's divisor problem. Linnik's creation of the dispersion method as an effective tool in the theory of numbers was an outstanding achievement; the method does, nevertheless, have the disadvantage of being complicated to apply, the length of [52] being seventy-eight pages. The error term in the formula, which was identical with that obtained in the earlier conditional treatment, was subsequently sharpened by Bredikhin [3], who made several other applications of the dispersion method.

In 1965 a new element in the situation was introduced by the publication of Bombieri's theorem on primes in arithmetic progressions [2]. Since this theorem in the context had a strength almost equivalent to what could be deduced about arithmetic progressions on Hypothesis R, the author's treatment now led at once to an alternative unconditional proof of the Hardy–Little-wood conjecture, only minor changes in [31] being necessary; some substantial comments on this application of Bombieri's theorem were made by Elliott and Halberstam [10] in 1966. This proof, albeit not the first to establish the result unconditionally, is undoubtedly shorter and easier than Linnik's.

Though we do not propose to be detained on the matter here, we should mention that Linnik and his school have also estab-lished results on a generalization of the problem in which the solutions of the equation

$$\phi(u, v) + p = n$$

are considered, $\phi(u, v)$ being a binary quadratic form. The proofs of some of these results can likewise now be simplified by using Bombieri's theorem.

Recently interest has been aroused in the associated problem of determining a formula for the number of primes not exceeding x that are of the form $u^2 + v^2 + a$, where the primes are not now to be counted according to the multiplicity of their representa-tion in the proposed manner. While no doubt this number is asymptotically equivalent to a fixed multiple of $x/\log^{\frac{3}{2}} x$, the problem of substantiating this may well lie very deep. Iwaniec

[47] has, nevertheless, recently proved that the *actual order* of magnitude of this number is indeed $x/\log^{\frac{3}{2}} x$ by using the $\frac{1}{2}$-residue sieve to obtain a lower bound, the corresponding upper bound having been familiar for a long time and being easily handled by the methods of Chapter 1. The relevance of the $\frac{1}{2}$-residue sieve to the problem stems from the consideration that it is enough to estimate from below the number of primes p not exceeding x for which $p-a$ is not divisible by primes congruent to 3, mod 4. In the light of the remarks made in Chapter 1, § 3, it follows that the sieving limit here is nearly $x^{\frac{1}{2}}$ if Bombieri's theorem is used; also p can be subjected to a simple extra congruential condition in order to ensure that $p-a$ cannot be divisible by just a single prime (without multiplicity) congruent to 3, mod 4. The problem is thus almost, but not quite, within the compass of a direct application of the $\frac{1}{2}$-residue sieve to the sequence $p-a$. Iwaniec, however, presses the matter to a successful conclusion by combining the $\frac{1}{2}$-residue sieve with another method in which the effect of p as a prime (as opposed to the effect of $p-a$ as a sum of two squares) is assessed through another sieve. A further proof is thus available for the infinitude of primes of the form u^2+v^2+a, and in much the same way it can be demonstrated that all sufficiently large numbers can be expressed as u^2+v^2+p. Yet the method is intricate and long.

We intend now to illustrate the use of the *enveloping sieve* by giving the shortest proof at present available of the Hardy–Littlewood conjecture. The treatment in the main is modelled on the author's original paper apart from the minor changes consequent upon the use of Bombieri's theorem. In the interests of brevity, Bredikhin's sharper error term is not obtained, but at the appropriate stage it is briefly indicated how the analysis can be refined in order to derive this improvement. We should end by pointing out that the upper bound sieve in its more conventional aspect is also required in the treatment.

2 Decomposition of sum

Let $\nu(n)$ be the number of representations of the integer n in the form

$$n = u^2 + v^2 + p.$$

Then

$$\nu(n) = \sum_{n=u^2+v^2+p} 1 = \sum_{p\leqslant n} \sum_{u^2+v^2=n-p} 1 = \sum_{p<n} r(n-p) + O(1). \quad (137)$$

Using the identity

$$r(\nu) = 4 \sum_{l|\nu} \chi(l),$$

where $\chi(l)$ is the non-principal character, mod 4, we also have

$$\sum_{p<n} r(n-p) = 4 \sum_{\substack{lm=n-p \\ p<n}} \chi(l)$$

$$= 4\Big(\sum_{l\leqslant n^{\frac{1}{2}}\log^{-48}n} + \sum_{n^{\frac{1}{2}}\log^{-48}n<l<n^{\frac{1}{2}}\log^{48}n} + \sum_{l\geqslant n^{\frac{1}{2}}\log^{48}n} \Big)$$

$$= 4(\Sigma_A + \Sigma_B + \Sigma_C), \quad \text{say}. \quad (138)$$

We have thus reduced the problem to that of the estimation of three different sums. It will transpire that Σ_A gives rise to the dominant term in the final asymptotic formula, Σ_B and Σ_C being of a lower order of magnitude.

3   Estimation of Σ_A and Σ_C

The estimations of Σ_A and Σ_C depend on some preliminary lemmata that we state and prove below. Here, as later in the chapter, it is assumed that $y \geqslant 1$.

Lemma 8 is a special case of Bombieri's theorem and can be inferred through partial summation from the form of the theorem as stated by Davenport [8]. Alternatively, provided minor consequential changes were made in our exposition, we could base our work on Gallagher's simpler treatment of the theorem [20].

LEMMA 8   *For $u \geqslant 2$ and $(a,k) = 1$, let*

$$E(u; a, k) = \pi(u; a, k) - \frac{\operatorname{li} u}{\phi(k)},$$

and define $E^*(x, k)$ by

$$E^*(x, k) = \operatorname*{bd\ max}_{2\leqslant u\leqslant x\ (a,k)=1} |E(u; a, k)|.$$

Then, for $y < 4x^{\frac{1}{2}}\log^{-48}x$, we have

$$\sum_{k\leqslant y} E^*(x, k) = O\Big(\frac{x}{\log^2 x}\Big).$$

LEMMA 9   *For any positive m we have*

$$\sum_{\substack{l \leqslant y \\ (l,\,m)=1}} \frac{\chi(l)}{l} = \frac{\pi}{4} \prod_{p|m} \left(1 - \frac{\chi(p)}{p}\right) + O\left(\frac{d(m;y)}{y}\right) + O\{\sigma_{-1}(m;y)\},$$

where $d(m;y) = \sum\limits_{\substack{d|m \\ d \leqslant y}} 1$ *and* $\sigma_{-1}(m;y) = \sum\limits_{\substack{d|m \\ d > y}} \frac{1}{d}$.

We have

$$\sum_{\substack{l \leqslant y \\ (l,\,m)=1}} \frac{\chi(l)}{l} = \sum_{l \leqslant y} \frac{\chi(l)}{l} \sum_{\substack{d|l \\ d|m}} \mu(d) = \sum_{\substack{d|m \\ d \leqslant y}} \frac{\mu(d)\,\chi(d)}{d} \sum_{t \leqslant y/d} \frac{\chi(t)}{t}$$

$$= \sum_{\substack{d|m \\ d \leqslant y}} \frac{\mu(d)\,\chi(d)}{d} \left\{\frac{\pi}{4} + O\left(\frac{d}{y}\right)\right\}$$

$$= \frac{\pi}{4} \sum_{d|m} \frac{\mu(d)\,\chi(d)}{d} + O\left(\frac{1}{y} \sum_{\substack{d|m \\ d \leqslant y}} 1\right) + O\left(\sum_{\substack{d|m \\ d > y}} \frac{1}{d}\right)$$

$$= \frac{\pi}{4} \prod_{p|m} \left(1 - \frac{\chi(p)}{p}\right) + O\left(\frac{d(m;y)}{y}\right) + O\{\sigma_{-1}(m;y)\},$$

as is asserted by the lemma.

LEMMA 10   *For any positive m we have*

$$\sum_{\substack{l \leqslant y \\ (l,\,m)=1}} \frac{\chi(l)}{\phi(l)} = \frac{\pi}{4} C \cdot E(m) + O\left(\frac{d(m)\log 2y}{y}\right),$$

where
$$C = \prod_{p>2} \left(1 + \frac{\chi(p)}{p(p-1)}\right)$$

and
$$E(m) = \prod_{\substack{p|m \\ p \equiv 1,\,\mathrm{mod}\,4}} \frac{(p-1)^2}{p^2-p+1} \prod_{\substack{p|m \\ p \equiv 3,\,\mathrm{mod}\,4}} \frac{p^2-1}{p^2-p-1}.$$

Also there exist positive absolute constants A_1, A_2 *such that*

$$A_1/\log\log 10m < E(m) < A_2 \log\log 10m.$$

Let l' denote a general square-free number. Then, since

$$\frac{l}{\phi(l)} = \prod_{p|l} \left(1 - \frac{1}{p}\right)^{-1} = \prod_{p|l} \left(1 + \frac{1}{p-1}\right) = \sum_{l'|l} \frac{1}{\phi(l')},$$

we have

$$\sum_{\substack{l \leqslant y \\ (l,\,m)=1}} \frac{\chi(l)}{\phi(l)} = \sum_{\substack{l \leqslant y \\ (l,\,m)=1}} \frac{\chi(l)}{l} \sum_{l'r=l} \frac{1}{\phi(l')} = \sum_{\substack{l' \leqslant y \\ (l',\,m)=1}} \frac{\chi(l')}{l'\phi(l')} \sum_{\substack{r \leqslant y/l' \\ (r,\,m)=1}} \frac{\chi(r)}{r}.$$

Therefore, the error terms in the statement of Lemma 9 being $O(d(m)/y)$, we obtain

$$\sum_{\substack{l \leqslant y \\ (l,\,m)=1}} \frac{\chi(l)}{\phi(l)} = \frac{\pi}{4} \prod_{p|m} \left(1 - \frac{\chi(p)}{p}\right) \sum_{\substack{l' \leqslant y \\ (l',\,m)=1}} \frac{\chi(l')}{l'\phi(l')} + O\left(\frac{d(m)}{y} \sum_{l' \leqslant y} \frac{1}{\phi(l')}\right)$$

$$= \frac{\pi}{4} \prod_{p|m} \left(1 - \frac{\chi(p)}{p}\right) \sum_{(l',\,m)=1} \frac{\chi(l')}{l'\phi(l')}$$

$$+ O\left\{\prod_{p|m} \left(1 + \frac{1}{p}\right) \sum_{l > y} \frac{1}{l\phi(l)}\right\} + O\left(\frac{d(m)}{y} \sum_{l \leqslant y} \frac{1}{\phi(l)}\right)$$

$$= \frac{\pi}{4} \prod_{p|m} \left(1 - \frac{\chi(p)}{p}\right) \sum_{(l',\,m)=1} \frac{\chi(l')}{l'\phi(l')} + O\left(\frac{d(m)\log 2y}{y}\right).$$

$$(139)$$

Furthermore, by Euler's principle,

$$\sum_{(l',\,m)=1} \frac{\chi(l')}{l'\phi(l')} = \prod_{\substack{p \nmid m \\ p > 2}} \left(1 + \frac{\chi(p)}{p(p-1)}\right) = C \prod_{\substack{p|m \\ p > 2}} \frac{p(p-1)}{p^2 - p + \chi(p)}, \quad (140)$$

and

$$\prod_{p|m} \left(1 - \frac{\chi(p)}{p}\right) \prod_{\substack{p|m \\ p > 2}} \frac{p(p-1)}{p^2 - p + \chi(p)} = \prod_{\substack{p|m \\ p > 2}} \frac{(p-1)(p - \chi(p))}{p^2 - p + \chi(p)} = E(m).$$

$$(141)$$

The first part of the lemma follows from (139), (140), and (141), the second part being an easy deduction from the inequality

$$\prod_{p|\nu} \left(1 - \frac{1}{p}\right)^{-1} = O(\log\log 10\nu).$$

We now consider Σ_A. We have, by (138),

$$\Sigma_A = \sum_{l \leqslant n^{\frac{1}{2}} \log^{-48} n} \chi(l) \sum_{\substack{p \equiv n,\, \mathrm{mod}\, l \\ p < n}} 1$$

$$= \sum_{\substack{l \leqslant n^{\frac{1}{2}} \log^{-48} n \\ (l,\,n)=1}} \chi(l) \sum_{\substack{p \equiv n,\, \mathrm{mod}\, l \\ p < n}} 1 + O\Big(\sum_{\substack{l \leqslant n^{\frac{1}{2}} \log^{-48} n \\ (l,\,n) > 1}} \sum_{\substack{p \equiv n,\, \mathrm{mod}\, l \\ p < n}} 1\Big)$$

$$= \sum_{\substack{l \leqslant n^{\frac{1}{2}} \log^{-48} n \\ (l,\,n)=1}} \chi(l) \sum_{\substack{p \equiv n,\, \mathrm{mod}\, l \\ p \leqslant n}} 1 + O\left(\frac{n^{\frac{1}{2}}}{\log^{48} n}\right),$$

since the arithmetic progression n, $\mathrm{mod}\, l$, contains at most one prime if $(l, n) > 1$. Hence, by Lemma 8,

$$\Sigma_A = \sum_{\substack{l \leqslant n^{\frac{1}{2}} \log^{-48} n \\ (l,\, n)=1}} \chi(l) \left(\frac{\operatorname{li} n}{\phi(l)} + E(n; n, l) \right) + O\left(\frac{n^{\frac{1}{2}}}{\log^{48} n} \right)$$

$$= \operatorname{li} n \sum_{\substack{l \leqslant n^{\frac{1}{2}} \log^{-48} n \\ (l,\, n)=1}} \frac{\chi(l)}{\phi(l)} + O\Big(\sum_{l \leqslant n^{\frac{1}{2}} \log^{-48} n} E^*(n, l) \Big) + O\left(\frac{n^{\frac{1}{2}}}{\log^{48} n} \right)$$

$$= \operatorname{li} n \sum_{\substack{l \leqslant n^{\frac{1}{2}} \log^{-48} n \\ (l,\, n)=1}} \frac{\chi(l)}{\phi(l)} + O\left(\frac{n}{\log^2 n} \right),$$

and so, by Lemma 10,

$$\Sigma_A = \frac{\pi}{4} C . E(n) \operatorname{li} n + O\left(\operatorname{li} n \frac{\log^{49} n}{n^{\frac{1}{2}}} d(n) \right) + O\left(\frac{n}{\log^2 n} \right)$$

$$= \frac{\pi}{4} C . E(n) \operatorname{li} n + O\left(\frac{n}{\log^2 n} \right)$$

$$= \frac{\pi}{4} C . E(n) \frac{n}{\log n} + O\left(\frac{n \log \log n}{\log^2 n} \right). \tag{142}$$

Next we consider Σ_C. In this the condition $l \geqslant n^{\frac{1}{2}} \log^{48} n$ implies that $m \leqslant n^{\frac{1}{2}} \log^{-48} n - 2n^{-\frac{1}{2}} \log^{-48} n = M_n$, say. Therefore

$$\Sigma_C = \sum_{m \leqslant M_n} \sum_{\substack{lm=n-p \\ lm \geqslant mn^{\frac{1}{2}} \log^{48} n}} \chi(l) = \sum_{m \leqslant M_n} \Sigma_{m, n}, \quad \text{say.} \tag{143}$$

The summand in $\Sigma_{m,n}$ is 1 if $l \equiv 1$, mod 4, is -1 if $l \equiv -1$, mod 4, and is 0 otherwise. Thus

$$\Sigma_{m, n} = \sum_{\substack{p \equiv n-m, \text{ mod } 4m \\ p \leqslant n - mn^{\frac{1}{2}} \log^{48} n}} 1 - \sum_{\substack{p \equiv n+m, \text{ mod } 4m \\ p \leqslant n - mn^{\frac{1}{2}} \log^{48} n}} 1.$$

Also the conditions $(n+m, 4m) = 1$ and $(n-m, 4m) = 1$ are equivalent. Hence, if $(n+m, 4m) = 1$, we may write

$$\Sigma_{m, n} = E(n - mn^{\frac{1}{2}} \log^{48} n; n-m, 4m)$$
$$\qquad\qquad - E(n - mn^{\frac{1}{2}} \log^{48} n; n+m, 4m)$$

$$= O(E^*(n, 4m)), \tag{144}$$

whereas, if $(n+m, 4m) > 1$, then trivially

$$\Sigma_{m, n} = O(1). \tag{145}$$

We thus have, by (143), (144), (145), and Lemma 8,

$$\Sigma_C = O(\sum_{m < n^{\frac{1}{4}} \log^{-48} n} \{E^*(n, 4m) + 1\}) = O\left(\frac{n}{\log^2 n}\right). \quad (146)$$

4   The enveloping sieve

We define the enveloping sieve that will be used in the estimation of Σ_B, and then develop as much of its machinery as our needs here dictate. We use the weaker variant of the Brun sieve that was described in Chapter 1, §1, in order to minimize the complications caused by our special requirements, the substitution of a stronger sieve having a negligible effect on the remainder term in the final result.

To define the sieve let

$$X = n^{\frac{1}{(\log \log n)^2}}, \quad P = \prod_{p \leqslant X} p,$$

and let d_1 indicate a typical divisor of P. Also, for any positive integer denoted by an arbitrary letter t, let $t^{(1)}$, $t^{(2)}$ be defined in terms of the representation

$$t = \prod p^\alpha$$

by

$$t^{(1)} = \prod_{p \leqslant X} p^\alpha, \quad t^{(2)} = \prod_{p > X} p^\alpha.$$

We define the function $f(\nu) = f_n(\nu)$ by

$$f(\nu) = g(\nu) + h(\nu),$$

where

$$g(\nu) = \begin{cases} 1, & \text{if } \nu \text{ is a prime not exceeding } X, \\ 0, & \text{otherwise}, \end{cases}$$

and

$$h(\nu) = \begin{cases} 1, & \text{if } (\nu, P) = 1, \\ 0, & \text{otherwise}. \end{cases}$$

Clearly $f(\nu)$ is a non-negative function that equals 1 when ν is a prime.

The discussion in Chapter 1, §1, immediately supplies a useful expression for $h(\nu)$ (take (6) with $S = (\nu)$). This is

$$h(\nu) = \sum_{\substack{d_1 | \nu \\ \omega(d_1) \leqslant r}} \mu(d_1) + O\left(\sum_{\substack{d_1 | \nu \\ \omega(d_1) = r+1}} 1 \right), \qquad (147)$$

where r is a given positive integer. We use this formula in the proof of Lemma 11, which should be contrasted with the Brun–Titchmarsh theorem as given by Lemma 1.

LEMMA 11 *Let $y \leqslant n$ and $k = O(n^\theta)$, where θ is a positive absolute constant less than 1. Then*

$$\sum_{\substack{\nu < y \\ \nu \equiv a, \, \text{mod} \, k}} f(\nu) = \begin{cases} \dfrac{1}{\phi(k)} B(n) \, y + O\left(\dfrac{n}{k \log^5 n} \right), & \text{if} \quad (a^{(1)}, k) = 1, \\[3mm] \quad O\left(\dfrac{n}{k \log^5 n} \right), & \text{if} \quad (a^{(1)}, k) > 1, \end{cases}$$

where $B(n)$ depends only on n and satisfies

$$B(n) = O\left(\frac{(\log \log n)^2}{\log n} \right).$$

In the above conditions $(a^{(1)}, k)$ may be replaced by $(a, k^{(1)})$.
Firstly, we have

$$\sum_{\substack{\nu < y \\ \nu \equiv a, \, \text{mod} \, k}} g(\nu) = O\left(\sum_{\nu \leqslant X} 1 \right) = O(X) = O\left(\frac{n}{k \log^5 n} \right). \qquad (148)$$

Secondly, suppose that $(a^{(1)}, k) = 1$. Then, by (147), we have

$$\sum_{\substack{\nu < y \\ \nu \equiv a, \, \text{mod} \, k}} h(\nu) = \sum_{\substack{\nu < y \\ \nu \equiv a, \, \text{mod} \, k}} \left\{ \sum_{\substack{d_1 | \nu \\ \omega(d_1) \leqslant r}} \mu(d_1) + O\left(\sum_{\substack{d_1 | \nu \\ \omega(d_1) = r+1}} 1 \right) \right\}$$

$$= \sum_{\omega(d_1) \leqslant r} \mu(d_1) \sum_\nu 1 + O\left(\sum_{\omega(d_1) = r+1} \sum_\nu 1 \right), \qquad (149)$$

where ν satisfies the conditions

$$\nu < y, \quad \nu \equiv a, \, \text{mod} \, k, \quad \nu \equiv 0, \, \text{mod} \, d_1,$$

in the inner summations on the right-hand side. Now the simultaneous congruences here have a solution in ν if and only if $(d_1, k) | a$, which condition is equivalent to $(d_1, k) = 1$ since $(a^{(1)}, k) = 1$. Furthermore, in this event, the solutions form a residue class, mod $d_1 k$. Therefore the right-hand side of (149) becomes

$$\sum_{\substack{(d_1,\, k)=1 \\ \omega(d_1)\leqslant r}} \mu(d_1)\left(\frac{y}{kd_1}+O(1)\right)+O\left\{\sum_{\substack{(d_1,\, k)=1 \\ \omega(d_1)=r+1}}\left(\frac{y}{kd_1}+O(1)\right)\right\}$$

$$=\frac{y}{k}\sum_{(d_1,\,k)=1}\frac{\mu(d_1)}{d_1}+O\left(\frac{y}{k}\sum_{\omega(d_1)>r}\frac{1}{d_1}\right)+O\left(\sum_{\omega(d_1)\leqslant r+1}1\right)$$

$$=\frac{y}{k}\Sigma_1+O\left(\frac{y}{k}\Sigma_2\right)+O(\Sigma_3),\quad\text{say.}\tag{150}$$

Now choosing $r=[10\log\log n]+1$, we have

$$\Sigma_1=\prod_{\substack{p\leqslant X \\ p\nmid k}}\left(1-\frac{1}{p}\right)=\prod_{p\leqslant X}\left(1-\frac{1}{p}\right)\prod_{p|k}\left(1-\frac{1}{p}\right)^{-1}\prod_{\substack{p|k \\ p>X}}\left(1-\frac{1}{p}\right)$$

$$=\frac{k}{\phi(k)}B(n)\prod_{\substack{p|k \\ p>X}}\left(1-\frac{1}{p}\right),\quad\text{say,}\tag{151}$$

where by the Mertens formula

$$B(n)\sim\frac{e^{-\gamma}}{\log X}=O\left(\frac{(\log\log n)^2}{\log n}\right).\tag{152}$$

But

$$\log\prod_{\substack{p|k \\ p>X}}\left(1-\frac{1}{p}\right)=\sum_{\substack{p|k \\ p>X}}\log\left(1-\frac{1}{p}\right)=O\left(\sum_{\substack{p|k \\ p>X}}\frac{1}{p}\right)=O\left(\frac{\log n}{X}\right),$$

since k has at most $\log_2 k$ different prime factors, and so

$$\prod_{\substack{p|k \\ p>X}}\left(1-\frac{1}{p}\right)=1+O\left(\frac{\log n}{X}\right)=1+O\left(\frac{1}{\log^5 n}\right).\tag{153}$$

Next we have

$$\Sigma_2\leqslant e^{-r}\sum_{d_1}\frac{e^{\omega(d_1)}}{d_1}=e^{-r}\prod_{p\leqslant X}\left(1+\frac{e}{p}\right)<e^{-r}\prod_{p\leqslant n}\left(1+\frac{1}{p}\right)^e=O(\log^{-a}n),$$

where $a=10-e>5$. Therefore

$$\Sigma_2=O\left(\frac{1}{\log^5 n}\right).\tag{154}$$

Furthermore, in Σ_3 we have

$$d_1\leqslant X^{\omega(d_1)}\leqslant n^{\frac{11}{\log\log n}}.$$

Therefore $\qquad\Sigma_3=O\left(n^{\frac{11}{\log\log n}}\right)=O\left(\frac{n}{k\log^5 n}\right).\tag{155}$

From (150), (151), (152), (153), (154), and (155) we deduce the first part of

$$\sum_{\substack{\nu \equiv a, \, \text{mod} \, k \\ \nu < y}} h(\nu) = \begin{cases} \dfrac{1}{\phi(k)} B(n) \, y + O\left(\dfrac{n}{k \log^5 n}\right), & \text{if} \quad (a^{(1)}, k) = 1, \\ 0, & \text{if} \quad (a^{(1)}, k) > 1, \end{cases}$$

(156)

the second part being trivial, since $(a^{(1)}, k) > 1$ and $\nu \equiv a$, mod k, implies $(\nu, P) > 1$.

The first part of the lemma follows from (148) and (156); the second part from $(a^{(1)}, k) = (a, k^{(1)})$.

5   Further lemmata

For convenience we collect together the other lemmata upon which the estimation of Σ_B depends.

We require a lemma concerning the distribution of numbers m for which the value of $\Omega(m)$ is restricted by certain conditions. Although a method due to Hardy and Ramanujan is applicable to problems of this nature [27], we prefer to use another method that has the advantage of requiring less calculation. As a preliminary we state

LEMMA 12   *If* $\frac{1}{2} \leqslant a \leqslant \frac{7}{4}$, *then*

$$\sum_{m \leqslant y} a^{\Omega(m)} = O(y \log^{a-1} 2y).$$

We merely indicate the proof, since it follows very closely that of a theorem due to Ramanujan and Wilson [83]. For $\sigma > 1$,

$$\sum_{m=1}^{\infty} \frac{a^{\Omega(m)}}{m^s} = \prod_p \left(1 - \frac{a}{p^s}\right)^{-1},$$

since the infinite product is absolutely convergent if $a < 2$. This product equals $\{\zeta(s)\}^a f(s)$, where $f(s)$ is given by a Dirichlet series that converges absolutely in a fixed half-plane of the form $\sigma > 1 - \delta$ ($\delta > 0$). Hence

$$\sum_{m \leqslant y} a^{\Omega(m)} = A(a) \, y \log^{a-1} y + O(y \log^{a-2} y),$$

from which the lemma follows.

Lemma 13   *If* $\frac{1}{2} \leqslant \alpha < 1 < \beta \leqslant \frac{3}{2}$ *and* $y > e^e$, *then*

(i) $\displaystyle\sum_{\substack{y^{\frac{1}{2}}\log^{-50}y < m < y^{\frac{1}{2}}\log^{48}y \\ \Omega(m) \leqslant \alpha \log\log y}} \frac{1}{m} = O(\log^{\gamma_\alpha - 1} y . \log\log y),$

(ii) $\displaystyle\sum_{\substack{m \leqslant y \\ \Omega(m) \geqslant \beta \log\log y - 1}} \frac{1}{m} = O(\log^{\gamma_\beta} y),$

where $\gamma_c = c - c \log c.$

If $\frac{1}{2} \leqslant a \leqslant 1$, then by Lemma 12 and partial summation,

$$\sum_{y^{\frac{1}{2}}\log^{-50}y < m < y^{\frac{1}{2}}\log^{48}y} \frac{a^{\Omega(m)}}{m} = O(\log^{a-1} y . \log\log y).$$

Hence

$$\sum_{\substack{y^{\frac{1}{2}}\log^{-50}y < m < y^{\frac{1}{2}}\log^{48}y \\ \Omega(m) \leqslant \alpha \log\log y}} \frac{1}{m} \leqslant a^{-\alpha \log\log y} \sum_{y^{\frac{1}{2}}\log^{-50}y < m < y^{\frac{1}{2}}\log^{48}y} \frac{a^{\Omega(m)}}{m}$$

$$= O(\log^{(\gamma\alpha - a - 1)} y . \log\log y), \qquad (157)$$

where $\gamma_{c,a} = a - c \log a.$

Similarly, if $1 \leqslant a \leqslant \frac{3}{2}$,

$$\sum_{\substack{m \leqslant y \\ \Omega(m) \geqslant \beta \log\log y - 1}} \frac{1}{m} \leqslant a^{1-\beta \log\log y} \sum_{m \leqslant y} \frac{a^{\Omega(m)}}{m} = O(\log^{\gamma_{\beta,a}} y). \quad (158)$$

Now, for c fixed, $\gamma_{c,a}$ attains its minimum $c - c \log c$ when $a = c$. The lemma is thus true, since we may legitimately set $a = \alpha$ and $a = \beta$ in (157) and (158), respectively.

Lemma 14   *If* $(rs, m) = 1$ *and* $r, s, m \leqslant n$, *then*

$$\sum_{\substack{l \geqslant y \\ (l, ms)=1}} \frac{\chi(l)}{\phi(rsl)} = O\{\log\log n . R_m(r; s; y)\}$$

$$+ O\left(\log\log n \frac{\sigma_{-1}(s)}{rs} \sigma_{-1}(m; y)\right) + O\left(\frac{(\log\log n)^2}{rsy}\right),$$

where $\displaystyle R_m(r; s; y) = \frac{\log 2y}{y} \frac{d(s)}{rs} d(m; y).$

The same estimate applies to the sum obtained by replacing
$l \geqslant y$ *by* $l > y$ *in the conditions of summation.*

The proof is similar to that of Lemma 9. Using the relation

$$\frac{1}{\phi(rsl)} = \frac{1}{l\phi(rs)} \prod_{\substack{p|l \\ p\nmid r}} \left(1 - \frac{1}{p}\right)^{-1}$$

that is valid for $(l, s) = 1$, we obtain the identity

$$\frac{1}{\phi(rsl)} = \frac{1}{\phi(rs)} \cdot \frac{1}{l} \sum_{\substack{q|l \\ (q,\, r)=1}} \frac{1}{\phi(q)},$$

where q denotes a general square-free number. Therefore

$$\phi(rs) \sum_{\substack{l \geqslant y \\ (l,\, ms)=1}} \frac{\chi(l)}{\phi(rsl)} = \sum_{\substack{qt \geqslant y \\ (q,\, rsm)=1 \\ (t,\, sm)=1}} \frac{\chi(qt)}{qt\phi(q)}$$

$$= \sum_{\substack{q \leqslant y \\ (q,\, rsm)=1}} \frac{\chi(q)}{q\phi(q)} \sum_{\substack{t \geqslant y/q \\ (t,\, sm)=1}} \frac{\chi(t)}{t} + \sum_{(t,\, sm)=1} \frac{\chi(t)}{t} \sum_{\substack{q > y \\ (q,\, rsm)=1}} \frac{\chi(q)}{q\phi(q)},$$

and this, by Lemma 9, is equal to

$$O\left\{\sum_{q \leqslant y} \frac{1}{q\phi(q)} \cdot \frac{q}{y} d\left(sm; \frac{y}{q}\right)\right\} + O\left\{\sum_{q \leqslant y} \frac{1}{q\phi(q)} \sigma_{-1}\left(sm; \frac{y}{q}\right)\right\}$$

$$+ O\left(\frac{\log \log n}{y}\right)$$

$$= O\left(\frac{\log 2y}{y} d(s) d(m; y)\right) + O\left\{\sum_{l \leqslant y} \frac{1}{l\phi(l)} \sigma_{-1}\left(sm; \frac{y}{l}\right)\right\}$$

$$+ O\left(\frac{\log \log n}{y}\right). \quad (159)$$

Now

$$\sum_{l \leqslant y} \frac{1}{l\phi(l)} \sigma_{-1}\left(sm; \frac{y}{l}\right) < \sum_{l=1}^{\infty} \frac{1}{l\phi(l)} \sum_{\substack{d|sm \\ d > y/l}} \frac{1}{d}$$

$$= \sum_{\substack{d|sm \\ d \leqslant y}} \frac{1}{d} \sum_{l > y/d} \frac{1}{l\phi(l)} + \sum_{\substack{d|sm \\ d > y}} \frac{1}{d} \sum_{l=1}^{\infty} \frac{1}{l\phi(l)}$$

$$= O\left(\frac{1}{y} d(sm; y)\right) + O\{\sigma_{-1}(sm; y)\}$$

$$= O\left(\frac{1}{y} d(s) d(m; y)\right) + O\{\sigma_{-1}(sm; y)\}. \quad (160)$$

We transform $\sigma_{-1}(sm; y)$. If $\mu | sm$, then $\mu = \mu_1\mu_2$, where $\mu_1 | s$ and $\mu_2 | m$. Therefore

$$\sigma_{-1}(sm:y) \leqslant \sum_{\mu_1 | s} \frac{1}{\mu_1} \sum_{\substack{\mu_2 | m \\ \mu_2 > y/\mu_1}} \frac{1}{\mu_2} = \sum_{\mu_1 | s} \frac{1}{\mu_1} \sum_{\substack{\mu_2 | m \\ y/\mu_1 < \mu_2 \leqslant y}} \frac{1}{\mu_2} + \sum_{\mu_1 | s} \frac{1}{\mu_1} \sum_{\substack{\mu_2 | m \\ \mu_2 > y}} \frac{1}{\mu_2}$$

$$\leqslant \frac{1}{y} \sum_{\mu_1 | s} \sum_{\substack{\mu_2 | m \\ y/\mu_1 < \mu_2 \leqslant y}} 1 + \sigma_{-1}(s)\,\sigma_{-1}(m; y)$$

$$\leqslant \frac{d(s)\,d(m; y)}{y} + \sigma_{-1}(s)\,\sigma_{-1}(m; y). \tag{161}$$

The main part of the lemma follows from (159), (160), (161), and the inequality $rs/\phi(rs) = O(\log \log n)$, the last part being then an almost trivial consequence.

LEMMA 15 *If* $1 < u < n$; $u' \geqslant u$; *and* $m \leqslant n$, *then we have*

(i) $\displaystyle \sum_{d \leqslant u} \sum_{u/d < l < (u\log^{96} n)/d} R_m\!\left(d; l; \frac{u'}{d}\right) = O\{(\log\log n)^4\}.$

(ii) $\displaystyle \sum_{d \leqslant u} \sum_{u/d < l < (u\log^{96} n)/d} \frac{\sigma_{-1}(l)}{dl} \sigma_{-1}\!\left(m; \frac{u'}{d}\right) = O\{(\log\log n)^3\}.$

(iii) $\displaystyle \sum_{d \leqslant u} \sum_{u/d < l < (u\log^{96} n)/d} \frac{d}{u}\frac{1}{dl} = O\{\log\log n\}.$

As the detailed proof of this lemma involves a routine application of principles that are not of direct interest to us in this tract, we are content to illustrate the method by sketching a proof of part (i).

The first sum is equal to

$$\sum_{d \leqslant u} \frac{d}{u'} \log\frac{2u'}{d} d\left(m; \frac{u'}{d}\right) \sum_{u/d < l < (u\log^{96} n)/d} \frac{d(l)}{dl}$$

$$= O\left\{\frac{1}{u'} \sum_{d \leqslant u} \log\frac{2u'}{d} d\left(m; \frac{u'}{d}\right) \left[\log\left(\frac{u\log^{96} n}{d}\right) \log\log n\right]\right\}$$

$$= O\left\{\frac{\log\log n}{u'} \sum_{d \leqslant u'} \log^2\frac{2u'}{d} d\left(m; \frac{u'}{d}\right)\right\}$$

$$\qquad\qquad + O\left\{\frac{(\log\log n)^2}{u'} \sum_{d \leqslant u'} \log\frac{2u'}{d} d\left(m; \frac{u'}{d}\right)\right\}. \tag{162}$$

Also, for $k = 1, 2$, we have

$$\sum_{d \leqslant u'} \log^k \frac{2u'}{d} d\left(m; \frac{u'}{d}\right) = \sum_{\substack{l|m \\ l \leqslant u'}} \sum_{d \leqslant u'/l} \log^k \frac{2u'}{d} = O\left(\sum_{\substack{l|m \\ l \leqslant u'}} \frac{u'}{l} \log^k 2l\right)$$

$$= O\left(u' \sum_{l|m} \frac{\log^k 2l}{l}\right). \quad (163)$$

To complete the proof we use the inequality

$$\sum_{l|m} \frac{\log^k 2l}{l} = O\{(\log \log 10m)^{k+1}\}, \quad (164)$$

the demonstration of which we illustrate by considering the case $k = 1$. We have here that

$$\sum_{l|m} \frac{\log 2l}{l} \leqslant \sum_{l|m} \frac{\log l}{l} + \sum_{l|m} \frac{1}{l} = \sum_{l|m} \frac{\log l}{l} + O(\log \log 10m). \quad (165)$$

In this

$$\sum_{l|m} \frac{\log l}{l} = \sum_{l|m} \frac{1}{l} \sum_{\rho | l} \Lambda(\rho) = \sum_{\rho | m} \frac{\Lambda(\rho)}{\rho} \sum_{r | m/\rho} \frac{1}{r} = O\left(\log \log 10m \sum_{\rho | m} \frac{\Lambda(\rho)}{\rho}\right)$$

$$= O\left(\log \log 10m \sum_{\rho \leqslant \log 10m} \frac{\Lambda(\rho)}{\rho}\right) + O\left(\frac{\log \log 10m}{\log 10m} \sum_{\rho | m} \Lambda(\rho)\right)$$

$$= O\{(\log \log 10m)^2\}. \quad (166)$$

The inequality for the special case $k = 1$ follows from (165) and (166), the demonstration for the other case $k = 2$ being based on a more complicated application of the above principles.

The first part of the lemma is obtained from (162), (163), and (164).

6 Inequality for Σ_B

We are now in a position to commence our assessment of Σ_B. Let

$$D(m) = \sum_{\substack{l|m \\ n^{\frac{1}{2}} \log^{-48} n < l < n^{\frac{1}{2}} \log^{48} n}} 1 \quad \text{and} \quad F(m) = \sum_{\substack{l|m \\ n^{\frac{1}{2}} \log^{-48} n < l < n^{\frac{1}{2}} \log^{48} n}} \chi(l).$$

Then

$$\Sigma_B = \sum_{p < n} F(n - p) = \sum_{\substack{p < n \\ D(n-p) \neq 0}} F(n - p).$$

Therefore, by the Cauchy–Schwarz inequality,

$$\Sigma_B = O\{(\sum_{\substack{p<n \\ D(n-p)\neq 0}} 1)^{\frac{1}{2}} (\sum_{p<n} F^2(n-p))^{\frac{1}{2}}\} = O\{(\Sigma_D)^{\frac{1}{2}} (\Sigma_E)^{\frac{1}{2}}\}, \quad (167)$$

say.

In the next sections we consider Σ_D and Σ_E separately. In each case the treatment depends on a sieve method, the estimation of Σ_D involving the use of Lemmata 1 and 2 and the estimation of Σ_E involving the use of the enveloping sieve and Lemma 11.

7   Estimation of Σ_D

Let α satisfy the inequality $1 < \alpha \leqslant \frac{3}{2}$. Then

$$\Sigma_D \leqslant \sum_{\substack{p<n \\ \Omega(n-p)\leqslant\alpha\log\log n}} D(n-p) + \sum_{\substack{p<n \\ \Omega(n-p)>\alpha\log\log n}} 1$$

$$= \Sigma_{D,1} + \Sigma_{D,2}, \quad \text{say.} \quad (168)$$

We estimate $\Sigma_{D,1}$ first. Since here, and also later, sums over complicated ranges of the variables will occur, we shall adopt an abbreviated notation for some of the more lengthy conditions of summation. For the estimations that follow we define (L), (M), (P) to be the conditions

$$n^{\frac{1}{2}}\log^{-48}n < l < n^{\frac{1}{2}}\log^{48}n, \quad n^{\frac{1}{2}}\log^{-50}n < m < n^{\frac{1}{2}}\log^{48}n, \quad lm = n-p,$$

respectively. If (P) holds, the conditions of summation in $\Sigma_{D,1}$ imply that at least one of $\Omega(l)$ and $\Omega(m)$ does not exceed $\frac{1}{2}\alpha\log\log n$. Therefore

$$\Sigma_{D,1} \leqslant \sum_{\substack{(L),\,(P) \\ \Omega(l)\leqslant\frac{1}{2}\alpha\log\log n}} 1 + \sum_{\substack{(L),\,(P) \\ \Omega(m)\leqslant\frac{1}{2}\alpha\log\log n}} 1. \quad (169)$$

The conditions of summation in the final sum imply that $m < n^{\frac{1}{2}}\log^{48}n$. Therefore

$$\Sigma_{D,1} = \sum_{\substack{(L),\,(P) \\ \Omega(l)\leqslant\frac{1}{2}\alpha\log\log n}} 1 + \sum_{\substack{(M),\,(P) \\ \Omega(m)\leqslant\frac{1}{2}\alpha\log\log n}} 1 + \sum_{\substack{l<n^{\frac{1}{2}}\log^{48}n \\ m\leqslant n^{\frac{1}{2}}\log^{-50}n}} 1$$

$$= \Sigma_1 + \Sigma_2 + \Sigma_3, \quad \text{say.}$$

Since in this it is evident that

$$\Sigma_1 = O(\Sigma_2)$$

and
$$\Sigma_3 = O\left(\frac{n}{\log^2 n}\right),$$
it follows that
$$\Sigma_{D,1} = O(\Sigma_2) + O\left(\frac{n}{\log^2 n}\right). \tag{170}$$

Next
$$\Sigma_2 = \sum_{\substack{(M) \\ \Omega(m) \leqslant \frac{1}{2}\alpha \log\log n}} \sum_{\substack{p \equiv n,\, \mathrm{mod}\, m \\ p < n}} 1 = O\left(\frac{n}{\log n} \sum_{\substack{(M) \\ \Omega(m) \leqslant \frac{1}{2}\alpha \log\log n}} \frac{1}{\phi(m)}\right)$$

by the Brun–Titchmarsh theorem in the form of Lemma 1. Therefore, by Lemma 13,

$$\Sigma_2 = O\left(\frac{n \log\log n}{\log n} \sum_{\substack{(M) \\ \Omega(m) \leqslant \frac{1}{2}\alpha \log\log n}} \frac{1}{m}\right)$$

$$= O\left(\frac{n}{\log n} \log^{(\gamma_{\frac{1}{2}\alpha}-1)} n \cdot (\log\log n)^2\right),$$

since $\frac{1}{2} < \frac{1}{2}\alpha \leqslant \frac{3}{4}$. As $\gamma_{\frac{1}{2}\alpha} > 0$ we deduce from this and (170) that

$$\Sigma_{D,1} = O\left(\frac{n}{\log n} \log^{(\gamma_{\frac{1}{2}\alpha}-1)} n \cdot (\log\log n)^2\right). \tag{171}$$

The estimation of $\Sigma_{D,2}$ is due essentially to Erdös [11]. We have

$$\Sigma_{D,2} \leqslant \sum_{\substack{m < n \\ \Omega(m) > 10 \log\log n}} 1 + \sum_{\substack{p < n \\ \alpha \log\log n < \Omega(n-p) \leqslant 10 \log\log n}} 1$$
$$= \Sigma_4 + \Sigma_5, \quad \text{say.} \quad (172)$$

In this, by Lemma 12,

$$\Sigma_4 < \frac{1}{\log^5 n} \sum_{m < n} (\sqrt{e})^{\Omega(m)} = O(n \log^{\sqrt{e}-6} n) = O\left(\frac{n}{\log^2 n}\right). \tag{173}$$

To estimate Σ_5 let R_n be the set of numbers m not exceeding n that have no prime factors exceeding $n^{1/20 \log\log n}$. Then

$$\Sigma_5 \leqslant \sum_{\substack{m \in R_n \\ \Omega(m) \leqslant 10 \log\log n}} 1 + \sum_{\substack{n-p \notin R_n \\ \Omega(n-p) > \alpha \log\log n}} 1 = \Sigma_6 + \Sigma_7, \quad \text{say.} \quad (174)$$

Now, if $\Omega(m) \leqslant 10 \log\log n$ and $m \in R_n$, then $m \leqslant n^{\frac{1}{2}}$. Therefore

$$\Sigma_6 = O(n^{\frac{1}{2}}). \tag{175}$$

H A O

If $n - p \notin R_n$ and $\Omega(n-p) > \alpha \log \log n$, then $n - p$ has at least one representation in the form rp', where $p' > n^{1/20 \log \log n}$ and $\Omega(r) > \alpha \log \log n - 1$; in such a representation

$$r < n^{(1 - 1/20 \log \log n)}.$$

Therefore

$$\Sigma_7 \leqslant \sum_{\substack{r < n^{1-1/20 \log \log n} \\ \Omega(r) > \alpha \log \log n - 1}} \sum_{\substack{p < n \\ n-p=rp'}} 1 = O\left(\sum_{\substack{r < n^{1-1/20 \log \log n} \\ \Omega(r) > \alpha \log \log n - 1}} \frac{n^2}{\phi(nr) \log^2(n/r)} \right)$$

by Lemma 2, and so

$$\Sigma_7 = O\left(\frac{n(\log \log n)^3}{\log^2 n} \sum_{\substack{r \leqslant n \\ \Omega(r) \geqslant \alpha \log \log n - 1}} \frac{1}{r} \right)$$

$$= O\left(\frac{n}{\log n} \log^{(\gamma_a - 1)} n (\log \log n)^3 \right) \tag{176}$$

by Lemma 13. Since $\gamma_\alpha > 0$, equations (172), (173), (174), (175), and (176) imply

$$\Sigma_{D,2} = O\left(\frac{n}{\log n} \log^{(\gamma_a - 1)} n (\log \log n)^3 \right). \tag{177}$$

If we choose α so that $\gamma_\alpha = \gamma_{\frac{1}{2}\alpha}$, then

$$\alpha - \alpha \log \alpha = \tfrac{1}{2}\alpha - \tfrac{1}{2}\alpha \log(\tfrac{1}{2}\alpha),$$

which gives $\log \alpha = 1 - \log 2$ and hence $\alpha = \tfrac{1}{2}e$, the condition $1 < \alpha \leqslant \tfrac{3}{2}$ being satisfied. Also $\gamma_\alpha = \tfrac{1}{2}e \log 2$. We therefore deduce from (168), (171), and (177) that

$$\Sigma_D = O\left(\frac{n}{\log n} \log^{-\gamma} n (\log \log n)^3 \right), \tag{178}$$

where $\gamma = 1 - \tfrac{1}{2}e \log 2$ (> 0).

The value of γ is susceptible to some improvement, since the condition $\Omega(n-p) \leqslant \alpha \log \log n$ in $\Sigma_{D,1}$ is only partially reflected in the condition that one or other of $\Omega(l)$, $\Omega(m)$ should not exceed $\tfrac{1}{2}\alpha \log \log n$ (see (169)). A more elaborate method was followed by Bredikhin [3], who treated a sum like Σ_2 that included the extra condition $\Omega(l) \leqslant \alpha \log \log n - \Omega(m)$. This yielded the estimate for $\Sigma_{D,1}$ with the exponent $\gamma_{\frac{1}{2}\alpha}$ replaced by $\gamma_\alpha - 1 + \alpha \log 2 + \epsilon$; the value of α chosen was therefore $1/\log 2$, and this gave rise to the improved value

$$\gamma = 1 - \frac{1}{\log 2} - \frac{\log \log 2}{\log 2} + \epsilon.$$

8 Estimation of Σ_E

Using the enveloping sieve function $f(\nu)$ defined in § 4, we have

$$\Sigma_E = \sum_{p<n} F^2(n-p) \leqslant \sum_{\nu<n} F^2(n-\nu)f(\nu)$$

$$= \sum_{\substack{l_1' m_1 = l_2' m_2 = n-\nu \\ n^{\frac{1}{2}} \log^{-48} n < l_1', l_2' < n^{\frac{1}{2}} \log^{48} n}} \chi(l_1') \chi(l_2') f(\nu).$$

Now, for given l_1', l_2', the values of $n-\nu$ in the above sum are congruent to 0, mod $[l_1', l_2']$. Furthermore, if $(l_1', l_2') = d$, then $l_1' = l_1 d$, $l_2' = l_2 d$, and $[l_1', l_2'] = dl_1 l_2$, where $(l_1, l_2) = 1$. We now define (L_i), (H), (K), (K_1) by

$$(L_i) \equiv \{(n^{\frac{1}{2}} \log^{-48} n)/d < l_i < (n^{\frac{1}{2}} \log^{48} n)/d\},$$

$$(H) \equiv \{(l_1, l_2) = 1\},$$

$$(K) \equiv \{(dl_1 l_2, n^{(1)}) = 1\},$$

$$(K_1) \equiv \{(dl_1, n^{(1)}) = 1\},$$

where $n^{(1)}$ is defined in terms of n by the notation described in § 4. Hence

$$\Sigma_E = \sum_{\substack{dl_1 l_2 m = n-\nu \\ (L_1),\, (L_2),\, (H)}} \chi^2(d)\, \chi(l_1)\, \chi(l_2) f(\nu) = \sum_{d \geqslant n^{\frac{1}{8}}} + \sum_{d < n^{\frac{1}{8}}} = \Sigma_1 + \Sigma_2, \quad \text{say.}$$

$$(179)$$

In Σ_1 we have $dl_1 l_2 < (n \log^{96} n)/d \leqslant n^{\frac{7}{8}} \log^{96} n$. Hence, by Lemma 11,

$$\Sigma_1 = \sum_{\substack{(L_1),\, (L_2),\, (H) \\ d \geqslant n^{\frac{1}{8}}}} \chi^2(d)\, \chi(l_1)\, \chi(l_2) \sum_{\substack{\nu \equiv n,\, \text{mod } dl_1 l_2 \\ \nu < n}} f(\nu)$$

$$= B(n)\, n \sum_{\substack{(L_1),\, (L_2),\, (H),\, (K) \\ d \geqslant n^{\frac{1}{8}}}} \frac{\chi^2(d)\, \chi(l_1)\, \chi(l_2)}{\phi(dl_1 l_2)}$$

$$+ O\left(\frac{n}{\log^5 n} \sum_{\substack{(L_1),\, (L_2),\, (H) \\ d \geqslant n^{\frac{1}{8}}}} \frac{1}{dl_1 l_2} \right)$$

$$= B(n)\, n \sum_{\substack{(L_1),\,(L_2),\,(H),\,(K) \\ n^{\frac{1}{8}} \leqslant d \leqslant n^{\frac{1}{2}} \log^{-48} n}} \frac{\chi^2(d)\,\chi(l_1)\,\chi(l_2)}{\phi(dl_1 l_2)}$$

$$+ B(n)\, n \sum_{\substack{(L_1),\,(L_2),\,(H),\,(K) \\ n^{\frac{1}{2}} \log^{-48} n < d < n^{\frac{1}{2}} \log^{48} n}} \frac{\chi^2(d)\,\chi(l_1)\,\chi(l_2)}{\phi(dl_1 l_2)}$$

$$+ O\left(\frac{n}{\log^5 n} \sum_{d,\,l_1,\,l_2 < n} \frac{1}{dl_1 l_2} \right)$$

$$= B(n)\, n\Sigma_3 + B(n)\, n\Sigma_4 + O\left(\frac{n}{\log^2 n} \right), \quad \text{say.} \tag{180}$$

We estimate Σ_3 through the equation

$$\Sigma_3 = \sum_{\substack{(L_1),\,(K_1) \\ n^{\frac{1}{8}} \leqslant d \leqslant n^{\frac{1}{2}} \log^{-48} n}} \chi^2(d)\,\chi(l_1) \sum_{\substack{(L_2) \\ (l_2,\,l_1 n^{(1)}) = 1}} \frac{\chi(l_2)}{\phi(dl_1 l_2)}.$$

Substituting for the inner sum by Lemma 14, and then deleting the conditions (K_1) and $d \geqslant n^{\frac{1}{8}}$ from the outer summation, we obtain

$$\Sigma_3 = O\left\{ \sum_{\substack{(L_1) \\ d \leqslant n^{\frac{1}{2}} \log^{-48} n}} \left(\log\log n \left[R_{n^{(1)}}\left(d; l_1; \frac{n^{\frac{1}{2}} \log^{-48} n}{d} \right) \right. \right. \right.$$

$$\left. + R_{n^{(1)}}\left(d; l_1; \frac{n^{\frac{1}{2}} \log^{48} n}{d} \right) \right]$$

$$+ \log\log n\, \frac{\sigma_{-1}(l_1)}{dl_1} \left[\sigma_{-1}\left(n^{(1)}; \frac{n^{\frac{1}{2}} \log^{-48} n}{d} \right) + \sigma_{-1}\left(n^{(1)}; \frac{n^{\frac{1}{2}} \log^{48} n}{d} \right) \right]$$

$$\left. \left. + (\log\log n)^2 \frac{d}{n^{\frac{1}{2}} \log^{-48} n} \cdot \frac{1}{dl_1} \right) \right\}.$$

If in this we set successively $m = n^{(1)}$, $u = n^{\frac{1}{2}} \log^{-48} n$, $u' = u$ and $m = n^{(1)}$, $u = n^{\frac{1}{2}} \log^{-48} n$, $u' = n^{\frac{1}{2}} \log^{48} n$, the conditions of Lemma 15 are satisfied. Therefore

$$\Sigma_3 = O\{(\log\log n)^5\} + O\{(\log\log n)^4\} + O\{(\log\log n)^3\}$$
$$= O\{(\log\log n)^5\}. \tag{181}$$

Also

$$\Sigma_4 = O\left(\sum_{\substack{n^{\frac{1}{2}} \log^{-48} n < d < n^{\frac{1}{2}} \log^{48} n \\ l_1,\, l_2 < \log^{96} n}} \frac{1}{\phi(dl_1 l_2)} \right)$$

$$= O\left(\log\log n \sum_{\substack{n^{\frac{1}{2}} \log^{-48} n < d < n^{\frac{1}{2}} \log^{48} n \\ l_1,\, l_2 < \log^{96} n}} \frac{1}{dl_1 l_2} \right) = O\{(\log\log n)^4\}. \tag{182}$$

We deduce from Lemma 11, (180), (181), and (182) that

$$\Sigma_1 = O\left(\frac{n}{\log n}(\log\log n)^7\right). \tag{183}$$

To estimate Σ_2 we use the fact that, for given l_1, l_2,

$$\sum_{\substack{rt=l_1\\st=l_2}}\mu(t) = \begin{cases} 1, & \text{if} \quad (l_1,l_2)=1,\\ 0, & \text{if} \quad (l_1,l_2)>1.\end{cases}$$

We define (R), (S), (D), (DT) by

$$(R) \equiv \{(n^{\frac{1}{2}}\log^{-48}n)/dt < r < (n^{\frac{1}{2}}\log^{48}n)/dt\},$$

(S) is the same as (R) except that s replaces r, $(D) \equiv \{d < n^{\frac{1}{8}}\}$, $(DT) \equiv \{d < n^{\frac{1}{8}}, t < n^{\frac{1}{8}}\}$. We thus have

$$\Sigma_2 = \sum_{\substack{rst^2dm=n-\nu\\(R),\,(S),\,(D)}} \mu(t)\,\chi^2(t)\,\chi^2(d)\,\chi(r)\,\chi(s)f(\nu)$$

$$= \sum_{t<n^{\frac{1}{8}}} + \sum_{t\geqslant n^{\frac{1}{8}}} = \Sigma_5 + \Sigma_6, \quad \text{say.} \tag{184}$$

As the conditions of summation in Σ_5 imply

$$rt^2dm < \frac{n}{s} < \frac{n}{n^{\frac{1}{2}}d^{-1}t^{-1}\log^{-48}n} = n^{\frac{1}{2}}dt\log^{48}n < n^{\frac{3}{4}}\log^{48}n, \tag{185}$$

we have

$$\Sigma_5 = \sum_{\substack{rt^2dm<n^{\frac{3}{4}}\log^{48}n\\(R),\,(DT)}} \mu(t)\,\chi^2(t)\,\chi^2(d)\,\chi(r)\sum_{\substack{\nu=n-rst^2dm\\(S)}}\chi(s)f(\nu)$$

$$= O\Bigl(\sum_{\substack{rt^2dm<n^{\frac{3}{4}}\log^{48}n\\(R),\,(DT)}} \Bigl|\sum_{\substack{\nu=n-rst^2dm\\y_1\leqslant\nu<y_2}}\chi(s)f(\nu)\Bigr|\Bigr), \tag{186}$$

where

$$y_2 = \max\,(n - n^{\frac{1}{2}}rtm\log^{-48}n, 1)$$

and

$$y_1 = \max\,([n - n^{\frac{1}{2}}rtm\log^{48}n] + 1, 1).$$

If we set $\lambda = rt^2dm$, we have that $\nu = n - \lambda s$. Also $\chi(s)$ is 1 if $s \equiv 1$, mod 4, is -1 if $s \equiv -1$, mod 4, and is 0 otherwise. Hence the inner sum in (186) is

$$\sum_{\substack{\nu\equiv n-\lambda,\,\text{mod }4\lambda\\y_1\leqslant\nu<y_2}} f(\nu) - \sum_{\substack{\nu\equiv n+\lambda,\,\text{mod }4\lambda\\y_1\leqslant\nu<y_2}} f(\nu), \tag{187}$$

where ν is the variable of summation. Now the conditions

$$(n - \lambda, 4\lambda^{(1)}) = 1 \quad \text{and} \quad (n + \lambda, 4\lambda^{(1)}) = 1$$

are equivalent, because each is equivalent to $(n - \lambda^{(1)}\lambda^{(2)}, 2\lambda^{(1)}) = 1$. Moreover, by (185), $4\lambda = O(n^{\frac{3}{4}} \log^{48} n)$. Therefore, by Lemma 11, the expression (187) is equal to

$$\begin{cases} \dfrac{y_2 - y_1}{\phi(4\lambda)} B(n) - \dfrac{y_2 - y_1}{\phi(4\lambda)} B(n) + O\left(\dfrac{n}{\lambda \log^5 n}\right) = O\left(\dfrac{n}{\lambda \log^5 n}\right), \\ \qquad\qquad\qquad\qquad\qquad\qquad \text{if} \quad (n - \lambda, 4\lambda^{(1)}) = 1, \\ O\left(\dfrac{n}{\lambda \log^5 n}\right), \quad \text{if} \quad (n - \lambda, 4\lambda^{(1)}) > 1. \end{cases}$$

Hence, by this and (186),

$$\Sigma_5 = O\left(\sum_{r,\,d,\,m,\,t^2 \leqslant n} \frac{n}{rdmt^2 \log^5 n}\right) = O\left(\frac{n}{\log^2 n}\right). \tag{188}$$

Also

$$\Sigma_6 = O\left(\sum_{\substack{rst^2dm \leqslant n \\ t \geqslant n^{\frac{1}{8}}}} 1\right) = O\left(\sum_{t \geqslant n^{\frac{1}{8}}} \sum_{\mu \leqslant n/t^2} d_4(\mu)\right)$$

$$= O\left(n \log^3 n \sum_{t \geqslant n^{\frac{1}{8}}} \frac{1}{t^2}\right) = O(n^{\frac{7}{8}} \log^3 n) = O\left(\frac{n}{\log^2 n}\right). \tag{189}$$

We infer from (184), (188), and (189) that

$$\Sigma_2 = O\left(\frac{n}{\log^2 n}\right). \tag{190}$$

Therefore, finally, by (179), (183), and (190), we have

$$\Sigma_E = O\left(\frac{n}{\log n} (\log \log n)^7\right). \tag{191}$$

9 The asymptotic formulae

The final results are now almost immediate. First from (167), (178), and (191) we have

$$\Sigma_B = O\left(\frac{n}{\log n} \log^{-\delta} n (\log \log n)^5\right),$$

where $\qquad\qquad \delta = \tfrac{1}{2}(1 - \tfrac{1}{2}e \log 2) \quad (> 0).$

The first part of the first theorem then follows from this, (137), (138), (142), and (146), the second part then being implied by the inequality for $E(m)$ in Lemma 10.

THEOREM 5 *The number of representations of the integer n in the form $n = p + u^2 + v^2$ is equal to*

$$\frac{\pi n}{\log n} \prod_{p>2} \left(1 + \frac{\chi(p)}{p(p-1)}\right) \prod_{\substack{p|n \\ p \equiv 1, \bmod 4}} \left(\frac{(p-1)^2}{p^2-p+1}\right) \prod_{\substack{p|n \\ p \equiv 3, \bmod 4}} \left(\frac{p^2-1}{p^2-p-1}\right)$$

$$+ O\left(\frac{n}{\log n} \log^{-\delta} n (\log\log n)^5\right).$$

Every sufficiently large number is a sum of a prime and two integral squares.

As stated earlier the conjugate theorem may be proved in a similar manner.

THEOREM 6 *We have, as $x \to \infty$,*

$$\sum_{0 < p+a \leqslant x} r(p+a) = \frac{\pi x}{\log x} \prod_{p>2} \left(1 + \frac{\chi(p)}{p(p-1)}\right)$$

$$\times \prod_{\substack{p|a \\ p \equiv 1, \bmod 4}} \left(\frac{(p-1)^2}{p^2-p+1}\right) \prod_{\substack{p|a \\ p \equiv 3, \bmod 4}} \left(\frac{p^2-1}{p^2-p-1}\right)$$

$$+ O\left(\frac{x}{\log x} \log^{-\delta} x (\log\log x)^5\right),$$

where a is a given non-zero integer.

There exist infinitely many primes of the form $u^2 + v^2 + a$.

We end by noting that the numerical value of the constant δ is approximately $\cdot 029$. This could be improved to approximately $\cdot 043$ by using Bredikhin's refinement that was described at the end of §7.

6. On the integers in an interval that are expressible as a sum of two squares – a new application of the lower bound sieve method

1 Introduction

The rôle of the sieve method in the solution of problems has characteristically been confined in the literature to the determination of some estimate or formula for the number of sifted elements in sequences that have some direct or indirect relevance to the end in view. To this pattern the applications of the sieve method made so far in this tract have been no exception, whatever other unconventional aspects they may have presented. We now shew by means of an example that sieve methods can in fact be utilized in other ways, even though, as in this example, the ultimate end in view – as opposed to the intermediate objective – may itself be an estimate for the number of sifted elements in a sequence.

To describe the background of the problem to be considered, let s denote, generally, a number that is expressible as a sum of two squares, and then let

$$M(x, h) = \sum_{x < s \leqslant x+h} 1.$$

Then, following the methods of Ingham, Montgomery, and Huxley [45] in connection with the similar problem for primes, it can be shewn that

$$M(x, h) \sim \frac{A_1 h}{\sqrt{(\log x)}}$$

as $x \to \infty$, provided that $x^\theta < h < x$ where $\theta > \frac{7}{12}$. For smaller values of h sieve methods can yield some comparable results in which asymptotic equality is replaced by upper or lower estimates of the same order of magnitude. This is a consequence of the property – already alluded to at the beginning of the previous chapter – that primes congruent to 3, mod 4, can enter into the

[98]

expression of a number s as a product of primes to even powers only, the sequence of the numbers s being very similar to the sequence obtained by eliminating all multiples of primes congruent to 3, mod 4, from the natural numbers. Thus the upper estimate

$$M(x,h) < \frac{A(\epsilon)\,h}{\sqrt{(\log x)}}$$

s almost immediate for $x^\epsilon < h < x$, while the $\frac{1}{2}$-residue sieve, to which reference was made in Chapter 1, §3, can be applied to obtain the corresponding lower bound for $x^{\frac{1}{2}+\epsilon} < h < x$, the sifting limit for the problem being $h^{1-\epsilon}$ ($> x^{\frac{1}{2}}$) (cf. remarks at the beginning of the previous chapter regarding the use of the $\frac{1}{2}$-residue sieve). The actual order of magnitude of $M(x,h)$ is therefore certainly $h/\sqrt{(\log x)}$ for $x^{\frac{1}{2}+\epsilon} < h < x$, and has been conjectured to be this for $x^\epsilon < h < x$ also.

We now describe a method whereby it can be shewn that

$$M(x,h) \asymp \frac{h}{\sqrt{(\log x)}}$$

in the extended range $x^{\frac{1}{3}+\epsilon} < h < x$, it being enough in the light of the above remarks to obtain the lower bound that is implicit in the statement. The treatment is based on a recent paper by the author [43] in which a similar but more complicated application of the method led to the conclusion that this order relation was actually valid in the slightly wider range $x^{\frac{12}{37}+\epsilon} < h < x$.

The genesis of the method lies in the remark that to any problem about the numbers s there often corresponds a parallel problem of lesser difficulty in which each number s is affected by a weight equal to the number of ways it can be represented as a sum of two squares. The problem analogous to the one at issue here thus concerns the sum

$$\sum_{x < n \leqslant x+h} r(n), \tag{192}$$

which is familiar in connection with the circle problem. The original sum is then approached through the sums derived from (192) by affecting $r(n)$ with a coefficient $t(n)$ that is bounded above by a multiple of $1/r(n)$ when $r(n) \neq 0$. The function $t(n)$ itself is defined in terms of functions that occur in the lower bound sieve method, although, unlike in conventional applications of sieve methods, there is no need whatsoever to attend closely to the choice of these functions in order to press home the method. Since $t(n)$ equals a sum over small divisors of n, the modified sums themselves can then be expressed in terms of sums S that are derived from (192) by subjecting the variable of summation therein to a divisibility condition. A final estimate can then flow from the properties of the sieving function and from asymptotic formulae that can be obtained for the sums S. Yet in practice, as in our working hereafter, it is usually convenient to depart from this procedure in some respects.

We end by making some brief comments on other applications of the method.

2 Lower bound for $2^{-\omega^*(m)}$

We develop the theory for the functions that provide lower bounds for the arithmetical function $2^{-\omega^*(m)}$, where $\omega^*(m)$ is the number of distinct prime factors of m that are congruent to 1, mod 4.

To formulate the method and to express its application, we introduce the convention that d, with or without subscript, indicates, generally, a positive square-free number (possibly 1) composed entirely of prime factors p such that $p \equiv 1$, mod 4. Next, for a given parameter v that later will be such that $v \to \infty$ as $x \to \infty$, we understand by δ or Δ, with or without subscript, a number of type d whose greatest prime factor does not exceed v. Also $\omega_v^*(m)$ is to be the number of distinct prime factors p of m such that $p \equiv 1$, mod 4, and $p \leqslant v$. Lastly $r_1(m)$ is defined to be 1 if m is a sum of two squares and to be 0 otherwise.

To define the lower bounding functions we use, as in the lower bound sieve discussed in Chapter 1, §3, a function $\rho(\delta) = \rho_{v,\xi}(\delta)$ with the property that

$$\rho(\delta) = 0 \tag{193}$$

for $\delta > \xi$ and with the property that

$$\sum_{\delta \mid m} \mu(\delta) \geqslant \sum_{\delta \mid m} \rho(\delta) \tag{194}$$

for all m. This function is considered in the first place in the general context of pairs of positive multiplicative functions $f(m)$, $f_1(\delta)$ that are characterized by the features:

(i) if $p \equiv 1$, mod 4, and $p \leqslant v$, then $0 < f(p) < 1$ and $f_1(p) = 1 - f(p)$;

(ii) otherwise, $f(p) = 1$;

(iii) in all cases $f(p^l) = f(p)$ for $l > 1$.

We have, for given Δ,

$$1 = \sum_{\delta | \Delta} \frac{f_1(\delta)}{f(\delta)} \sum_{\delta_1 | \delta} \mu(\delta_1) \geqslant \sum_{\delta | \Delta} \frac{f_1(\delta)}{f(\delta)} \sum_{\delta_1 | \delta} \rho(\delta_1)$$

$$= \sum_{\delta_1 \delta_2 \delta_3 = \Delta} \frac{f_1(\delta_1 \delta_2)}{f(\delta_1 \delta_2)} \rho(\delta_1) = \sum_{\delta_1 | \Delta} \frac{f_1(\delta_1) \rho(\delta_1)}{f(\delta_1)} \sum_{\delta_2 | \Delta / \delta_1} \frac{f_1(\delta_2)}{f(\delta_2)},$$

the inner sum in the last expression being

$$\prod_{p | \Delta / \delta_1} \left(1 + \frac{f_1(p)}{f(p)}\right) = \prod_{p | \Delta / \delta_1} \frac{1}{f(p)} = \frac{f(\delta_1)}{f(\Delta)}.$$

Therefore

$$f(\Delta) \geqslant \sum_{\delta_1 | \Delta} f_1(\delta_1) \rho(\delta_1),$$

and so, furthermore, for any integer m,

$$f(m) \geqslant \sum_{\delta | m} f_1(\delta) \rho(\delta).$$

We can obtain at once from this the most natural bound

$$\frac{1}{2^{\omega_v^*(m)}} \geqslant \sum_{\delta | m} \frac{\rho(\delta)}{2^{\omega(\delta)}}$$

on taking $f(p) = \tfrac{1}{2}$ for $p \equiv 1$, mod 4, and $p \leqslant v$. This, though it would suffice, is not convenient to use here, and we therefore select an alternative that is particularly suitable for our application. To find this we define $f(p)$ by

$$f(p) = 1 - \frac{1}{2}\left(1 - \frac{1}{p}\right)^{-1} \quad (< \tfrac{1}{2})$$

for $p \equiv 1$, mod 4, and $p \leqslant v$. This gives the inequality

$$\frac{1}{2^{\omega_v^*(m)}} \geqslant \sum_{\delta | m} \frac{\rho(\delta)}{2^{\omega(\delta)}} \frac{\delta}{\phi(\delta)}. \tag{195}$$

The above inequality for $2^{-\omega_v^*(m)}$ leads to a lower bound for $r_1(m)$ through the introduction of the function

$$r_2(m) = 4r_1(m)\, 2^{\omega^*(m)}$$

that is related to $r(m)$ by means of the identity

$$r_2(m) = \sum_{d^2\mid m} \mu(d)\, r(m/d^2). \tag{196}$$

This formula, which is analogous to the well-known

$$2^{\omega(m)} = \sum_{l^2\mid m} \mu(l)\, d(m/l^2),$$

is easily established, since one quarter of each side is multiplicative and since the case when m is a prime-power is almost immediate. From the definition of $r_2(m)$ we obtain

$$r_1(m) \geqslant \tfrac{1}{4} r_2(m)\, 2^{-\omega_v^*(m)}\, 2^{-\log m/\log v},$$

the number m having not more than $\log m/\log v$ prime factors exceeding v. Since in the sequel it will be assumed that $v > m^\beta$ for some $\beta > 0$, we deduce that we may apply the inequality in the form

$$r_1(m) \geqslant A(\beta)\, r_2(m)\, 2^{-\omega_v^*(m)}. \tag{197}$$

To discuss the form of $\rho(\delta)$ to be used in conjunction with the above formulae we must anticipate the last part of the chapter by stating that $\rho(\delta)$ must be taken so that the sum

$$\sum_\delta \frac{\rho(\delta)}{\delta}$$

is sufficiently large. This sum has, however, already been discussed in Chapter 1, §3, and what we need is easily brought together in the form of the following lemma.

LEMMA 16   *There exists a function* $\rho(\delta) = \rho_{v,\xi}(\delta)$ *with properties* (193), (194), *and with the properties*

  (i) $\rho(\delta) = O(\delta^\epsilon)$

  (ii) *there exists a positive absolute constant* $a < 1$ *such that*

$$\sum \frac{\rho(\delta)}{\delta} > \frac{A_2}{\sqrt{(\log v)}}$$

for $v = \xi^a$ $(v > v_0)$.

3 Lemmata

We prepare for the application of formula (197) by considering in succession a number of lemmata that are related to the circle problem.

LEMMA 17   *We have, for $h < y$,*

$$\sum_{y < m \leqslant y+h} r(m) = \pi h + O(y^{\frac{1}{4}}).$$

This is an immediate corollary of a classical result on the circle problem due to Sierpiński [73]. Later work has in fact shewn that the exponent of y in the remainder term can be slightly reduced, the value $\frac{12}{37} + \epsilon$ given by Chen Jing-run [7] being the best that is currently known. At the other extreme it is easily inferred from well-known Ω results on the circle problem that the lemma is not true without qualification for the exponent $\frac{1}{4}$. This does not, however, preclude the possibility that significant asymptotic formulae for the sum may be valid for values of h considerably less than $y^{\frac{1}{4}}$.

These remarks will be of importance in the final section when we consider possible extensions of the main theorem obtained in this chapter.

Next we apply the lemma to derive a similar formula in which $r(n)$ is replaced by $r_2(n)$.

LEMMA 18†   *Let $0 < \alpha < \frac{2}{3}$. Then, for $y^{\frac{1}{3}+\alpha} < h < y$, we have*

$$\sum_{y < m \leqslant y+h} r_2(m) = \pi B_1 h + O(hy^{-\frac{1}{4}\alpha}),$$

where
$$B_1 = \prod_{p \equiv 1, \bmod 4} \left(1 - \frac{1}{p^2}\right) > 0.$$

By (196), we have

$$\sum_{y < m \leqslant y+h} r_2(m) = \sum_{y < m \leqslant y+h} \sum_{d^2 l = m} \mu(d)\, r(l)$$

$$= \sum_{y < m \leqslant y+h} \left(\sum_{d \leqslant \xi_1} + \sum_{\xi_1 < d \leqslant \xi_2} + \sum_{d > \xi_2} \right) = \Sigma_1 + \Sigma_2 + \Sigma_3,$$

$$\text{say,} \quad (198)$$

where ξ_1 and ξ_2 are to be determined later so that $\xi_1 < \xi_2 < y^{\frac{1}{2}}$.

† Constant implied by O-notation depends at most on α. In the proof the constants implicit in the O-notation first depend on ϵ only and then on α only.

In this, by Lemma 17,

$$\Sigma_1 = \sum_{d \leqslant \xi_1} \mu(d) \sum_{y/d^2 < l \leqslant y/d^2 + h/d^2} r(l) = \pi h \sum_{d \leqslant \xi_1} \frac{\mu(d)}{d^2} + O\left(y^{\frac{1}{3}} \sum_{d \leqslant \xi_1} 1\right)$$

$$= \pi h \sum_{d} \frac{\mu(d)}{d^2} + O\left(\frac{h}{\xi_1}\right) + O(y^{\frac{1}{3}}\xi_1)$$

$$= \pi B_1 h + O\left(\frac{h}{\xi_1}\right) + O(y^{\frac{1}{3}}\xi_1). \tag{199}$$

Next, since $r(l) = O(l^\epsilon)$,

$$\Sigma_2 = O(y^\epsilon \sum_{\xi_1 < d \leqslant \xi_2} \sum_{y/d^2 < l \leqslant y/d^2 + h/d^2} 1)$$

$$= O\left\{y^\epsilon \sum_{\xi_1 < d \leqslant \xi_2} \left(\frac{h}{d^2} + 1\right)\right\}$$

$$= O\left(y^\epsilon h \sum_{d > \xi_1} \frac{1}{d^2}\right) + O\left(y^\epsilon \sum_{d \leqslant \xi_2} 1\right) = O\left(\frac{y^\epsilon h}{\xi_1}\right) + O(y^\epsilon \xi_2). \tag{200}$$

Finally, in Σ_3 the condition $d > \xi_2$ implies that $l \leqslant 2y/\xi_2^2$. Therefore

$$\Sigma_3 = O(y^\epsilon \sum_{l \leqslant 2y/\xi_2^2} \sum_{y/l < d^2 \leqslant y/l + h/l} 1)$$

$$= O\left(y^\epsilon \sum_{l \leqslant 2y/\xi_2^2} \left\{\left(\frac{y}{l} + \frac{h}{l}\right)^{\frac{1}{2}} - \left(\frac{y}{l}\right)^{\frac{1}{2}} + O(1)\right\}\right)$$

$$= O\left(y^\epsilon \sum_{l \leqslant 2y/\xi_2^2} \frac{h}{l^{\frac{1}{2}} y^{\frac{1}{2}}}\right) + O(y^\epsilon \sum_{l \leqslant 2y/\xi_2^2} 1)$$

$$= O\left(\frac{y^\epsilon h}{\xi_2}\right) + O\left(\frac{y^{1+\epsilon}}{\xi_2^2}\right) = O\left(\frac{y^\epsilon h}{\xi_1}\right) + O\left(\frac{y^{1+\epsilon}}{\xi_2^2}\right). \tag{201}$$

Combining the error terms in (199), (200), and (201) we obtain an asymptotic formula for the sum with remainder

$$O\left(\frac{y^\epsilon h}{\xi_1}\right) + O(y^{\frac{1}{3}}\xi_1) + O(y^\epsilon \xi_2) + O\left(\frac{y^{1+\epsilon}}{\xi_2^2}\right) = O(y^{\frac{1}{6}+\epsilon}h^{\frac{1}{2}}) + O(y^{\frac{1}{3}+\epsilon}),$$

on choosing $\xi_1 = y^{-\frac{1}{6}}h^{\frac{1}{2}}$ and $\xi_2 = y^{\frac{1}{3}}$ ($> \xi_1$). This in turn being

$$O(hy^{-\frac{1}{4}\alpha})$$

if $\epsilon = \frac{1}{4}\alpha$, we deduce that (198), (199), (200), and (201) give the lemma.

The last lemma in the series is

LEMMA 19 *Let* $0 < \alpha < \frac{2}{3}$. *Then, for* $y^{\frac{1}{3}+\alpha} < h < y$ *and for* $d \leqslant y^{\frac{1}{4}\alpha}$, *we have*

$$\sum_{y < dm \leqslant y+h} r_2(dm) = \frac{\pi B_1 h 2^{\omega(d)} \phi(d)}{d^2} + O\left(\frac{hy^{-\frac{1}{17}\alpha}}{d}\right).$$

Using the identity

$$r_2(dm) = 2^{\omega(d)} \sum_{\substack{d^*|d \\ d^*|m}} \mu(d^*) r_2(m/d^*),$$

the validity of which may be verified in a number of ways, we have

$$\sum_{y < dm \leqslant y+h} r_2(dm) = 2^{\omega(d)} \sum_{y < dm \leqslant y+h} \sum_{\substack{d^*|d \\ d^*|m}} \mu(d^*) r_2(m/d^*)$$

$$= 2^{\omega(d)} \sum_{d^*|d} \mu(d^*) \sum_{y/dd^* < l \leqslant y/dd^* + h/dd^*} r_2(l).$$

Since $(y/dd^*)^{\frac{1}{3}+\frac{1}{2}\alpha} \leqslant y^{\frac{1}{3}+\frac{1}{2}\alpha} < h/dd^* < y/dd^*$, we may apply Lemma 18 to the inner sum but with $\frac{1}{2}\alpha$ in place of α. Therefore

$$\sum_{y < dm \leqslant y+h} r_2(dm) = \frac{2^{\omega(d)}\pi B_1 h}{d} \sum_{d^*|d} \frac{\mu(d^*)}{d^*} + O\left(\frac{h 2^{\omega(d)} y^{-\frac{1}{16}\alpha}}{d} \sum_{d^*|d} \frac{1}{d^*}\right),$$

where the inequality $(y/dd^*)^{\frac{1}{8}\alpha} > y^{\frac{1}{16}\alpha}$ is used to estimate the error term. Consequently

$$\sum_{y < dm \leqslant y+h} r_2(dm) = \frac{\pi B_1 h 2^{\omega(d)} \phi(d)}{d^2} + O\left(\frac{hy^{-\frac{1}{17}\alpha}}{d}\right),$$

as required.

4 The order of $M(x, h)$

The lower bound for $M(x, h)$ is now quickly obtained. As before we let $0 < \alpha < \frac{2}{3}$ and take h so that $x^{\frac{1}{3}+\alpha} < h < x$; let also $\xi = x^{\frac{1}{4}\alpha}$

and $v = x^{\frac{1}{4}a\alpha}$, where a is the constant defined in Lemma 16. Then, by (197) followed by (195),

$$\sum_{x<n\leqslant x+h} r_1(n) \geqslant A_1(\alpha) \sum_{x<n\leqslant x+h} r_2(n)\, 2^{-\omega_v^*(n)}$$

$$\geqslant A_1(\alpha) \sum_{x<n\leqslant x+h} r_2(n) \sum_{\delta|n} \frac{\rho(\delta)}{2^{\omega(\delta)}} \frac{\delta}{\phi(\delta)}$$

$$= A_1(\alpha) \sum_{\delta} \frac{\rho(\delta)}{2^{\omega(\delta)}} \frac{\delta}{\phi(\delta)} \sum_{x<\delta m \leqslant x+h} r_2(\delta m),$$

from which and Lemmata 19 and 16 we infer that

$$\sum_{x<n\leqslant x+h} r_1(n) \geqslant A_2(\alpha)\, h \sum_{\delta} \frac{\rho(\delta)}{\delta} + O\left(hx^{-\frac{1}{18}\alpha} \sum_{\delta \leqslant \xi} \frac{1}{\delta}\right)$$

$$> \frac{A_3(\alpha)\, h}{\sqrt{(\log v)}} + O\left(\frac{h\log x}{x^{\frac{1}{18}\alpha}}\right)$$

$$> \frac{A_4(\alpha)\, h}{\sqrt{(\log x)}} \tag{202}$$

for $x > x_0(\alpha)$.

We thus have the following theorem on account of the remarks made in the introduction about the upper bound corresponding to (202).

Theorem 7 *Let $\frac{1}{3} < \theta < 1$, and let $M(x,h)$ be the number of integers in the interval $x < n \leqslant x + h$ that are expressible as the sum of two squares. Then*

$$M(x,h) \asymp \frac{h}{\sqrt{(\log x)}}$$

for $x^\theta < h < x$.

5 Other applications of the method

The result of Theorem 7 can in fact be shewn to hold in respect of any positive constant θ that has the property that

$$\sum_{y<n\leqslant y+h} r(n) = \pi h + O(y^\Theta) \tag{203}$$

for some constant $\Theta < \theta$. The incidental details in the proof, however, are harder for $\frac{1}{4} < \theta \leqslant \frac{1}{3}$ than in the case treated above. It will thus be observed that we can replace the lower bound $\frac{1}{3}$ for

θ by $\frac{12}{37}$ in view of Chen Jing-run's result, any improvement in the exponent of y in the remainder term of (203) leading to a corresponding improvement in our theorem. We should add, moreover, that still further improvements in the theorem would follow from the truth of significant asymptotic formulae for the sum in (203) for the case where h was small compared with $y^{\frac{1}{4}}$.

With appropriate modifications, the method described here has other applications to the theory of numbers. It can be used, for instance, to shew that the number of pairs s_1, s_2 satisfying

$$s_2 - s_1 = k, \quad s_1 \leqslant x,$$

has actual order of magnitude $x/\log x$. This also is a new result [43]. The lower bound inherent in this should be compared with the weaker lower bound of order $x/\log^{2\log 2 + \epsilon}x$ that has been obtained by Schwarz [67] for the case $k = 1$.

In some problems it is enough to weight $r(n)$ with a factor $t(n)$ that is merely a formal substitute for $2^{-\omega^*(n)}$. In such cases factors simpler than the one used in this chapter can be adopted, particularly if it is permissible to choose them so that they are not necessarily always positive. An example of such a factor is

$$\sum_{\substack{d \mid n \\ d < w}} \frac{\mu(d)}{2^{\omega(d)}} \left(1 - \frac{\log d}{\log w}\right),$$

which may be expected in some sense to mimic $2^{-\omega^*(n)}$. Used in the appropriate way these factors smooth out the fluctuations due to the function $r(n)$ and enable the numbers s to be considered so that those with an abnormally large number of prime factors do not have an untoward influence.

Such a factor was recently used by the author [39] to shew that

$$\sum_{s_{n+1} \leqslant x} (s_{n+1} - s_n)^\gamma = O(x \log^{\frac{1}{2}(\gamma-1)} x)$$

for $\gamma < \frac{5}{3}$, where $s_1, s_2, ..., s_n, ...$ are the numbers s in ascending size. The latter result should be compared with the inequality

$$\sum_{p_{n+1} \leqslant x} \frac{(p_{n+1} - p_n)^2}{p_n} = O(x \log^3 x)$$

that has been obtained by Selberg on the Riemann hypothesis [68].

7. Primes in sparse sequences – other applications of sieve methods

1 Introduction

Although the famous assertion of Fermat to the effect that all numbers of the form $2^{2^m} + 1$ are prime was disproved as long ago as 1732 by Euler, it still remains a complete mystery as to whether all but a finite number of them be primes, or all but a finite number be composite, or neither. Much the same position obtains for the Mersenne numbers $2^p - 1$. These are but the two most familiar examples of the many sparsely distributed sequences whose multiplicative structure is as yet almost entirely unknown.

Since sieve methods have yielded upper bounds of the expected true order of magnitude for the number of primes in sequences given by polynomials by using the ideas of Chapter 1, §2, it is natural to consider their effect on other sequences such as that of the Fermat numbers. It is, however, too much to hope that they would be successful in shewing that any such sequence contained infinitely many primes in view of their failure to substantiate similar conclusions for polynomial sequences. On the other hand it may not be unreasonable to expect that in some cases a sieve method might establish that almost all members of such a sequence were composite.

Yet a moment's consideration leads to the conclusion that even the field of application of upper bound sieve methods is necessarily circumscribed. To take the case of the Fermat numbers, if a Fermat number is divisible by a prime, then it is the only Fermat number divisible by that prime, the set of such prime divisors being as yet unknown (it might indeed be to all intents and purposes the sequence itself). There is thus no way in which the machinery of the sieve method can be applied, the very reasons that might lead one to believe in a modified form of Fermat's assertion being the ones that preclude the use of a sieve method.

In like manner the method is inapplicable to the Mersenne numbers.

Both the sequences so far considered are somewhat exceptional in that they are derived from special sequences of the form $a^n + b$ by eliminating those values of n for which the numbers are obviously composite. When $b \neq \pm 1$ the situation is different, and there is no need to restrict the values of n in order to secure a meaningful problem. Such sequences may be regarded as being next in point of difficulty after polynomial sequences, and we therefore shall consider the relevance of upper bound sieve methods to them. Later on we shall then consider another similar naturally generated sparse sequence. Our methods are only attended with a qualified success, and therefore our exposition is in part somewhat speculative in nature.

2   Primes in the sequence $a^n + b$

In what follows we assume for simplicity that $a = 2$, and then naturally that b is odd and not equal to ± 1.

First there are infinitely many composite numbers in the sequence. This is merely a special case of a generalization due to Morgan Ward [29] of Goldbach's well-known theorem to the effect that a polynomial cannot always take prime values; in fact, setting $f(n) = 2^n + b$, let $f(\nu) \neq 0, 1$ and let $p \mid f(\nu)$; then $p \mid f(\nu + r(p-1))$ by Fermat's theorem, and we see, moreover, that there is a positive density of numbers n for which $f(n)$ is composite.

Next we wish to determine whether almost all members of the sequence are composite. We therefore need to use a sequence of primes p such that $\Sigma \dfrac{1}{p}$ is divergent in order to sieve the sequence. Here at once the first difficulty arises because of our inability to find such a set of primes p that have the property that the congruence

$$2^\nu + b \equiv 0, \quad \mod p, \tag{204}$$

is soluble. This difficulty would, however, be removed if we knew the truth of the Artin conjecture which was proved in Chapter 3 under the assumption of the extended Riemann hypothesis. In this event the primes p for which 2 is a primitive root, $\mathrm{mod}\, p$, would satisfy

$$\sum_{p\leqslant x} \frac{1}{p} \sim C \log\log x,$$

where

$$C = \prod_q \left(1 - \frac{1}{q(q-1)}\right),$$

and for all such p not dividing b the congruence would be soluble, the solutions forming a single residue class, $\mathrm{mod}\,(p-1)$. A yet more serious difficulty then emerges because the usual sieve methods demand a knowledge of the solubility of the congruence

$$2^\nu + b \equiv 0, \quad \mathrm{mod}\, l, \tag{205}$$

for composite square-free numbers l formed with prime factors p from the set just described. However, the solubility of (204) for different primes $p_1, \ldots, p_r$ does not imply the solubility of (205) for $l = p_1 \ldots p_r$, since $p_1 - 1, \ldots, p_r - 1$ are not co-prime and may, indeed, have common factors other than 2. In fact there is no satisfactory formula for the number of solutions of (205) satisfying $\nu \leqslant x$ even if l is restricted to be a product of the primes of special type.

A way of minimizing the second difficulty is to adopt the procedure suggested in Chapter 1, § 6. We let $X = x^{\frac{1}{2}}$ and define $\omega_X(l)$ by

$$\omega_X(l) = \sum_{\substack{p \mid l \\ p \leqslant X}} 1,$$

where, until indicated otherwise, p, q in this section will denote primes not dividing b for which 2 is a primitive root, modulis p and q. We then consider the sums

$$\sum_{n\leqslant x} \omega_X(2^n + b), \tag{206}$$

$$\sum_{n\leqslant x} \omega_X^2(2^n + b) \tag{207}$$

with the intention of trying to compare their orders of magnitude.

The estimation of (206) presents no difficulty in principle. First

$$\sum_{n\leqslant x} \omega_X(2^n + b) = \sum_{n\leqslant x} \sum_{\substack{p \mid 2^n + b \\ p \leqslant X}} 1 = \sum_{p \leqslant X} \sum_{\substack{2^n + b \equiv 0,\, \mathrm{mod}\, p \\ n \leqslant x}} 1.$$

Next, for each p, the solutions of 2^n+b form one residue class, $\mod (p-1)$. Hence

$$\sum_{n\leqslant x} \omega_X(2^n+b) = \sum_{p\leqslant X}\left(\frac{x}{p-1}+O(1)\right)$$

$$= x\sum_{p\leqslant X}\frac{1}{p-1}+O(X) \sim Cx\log\log X. \qquad (208)$$

The estimation of (207) also proceeds along familiar lines in the first place. We have

$$\sum_{n\leqslant x} \omega_X^2(2^n+b) = \sum_{\substack{n\leqslant x \\ p\mid 2^n+b \\ q\mid 2^n+b \\ p,\,q\leqslant X}} 1 = \sum_{\substack{p,\,q\leqslant X}} \sum_{\substack{2^n+b\equiv 0,\,\mathrm{mod}\,p \\ 2^n+b\equiv 0,\,\mathrm{mod}\,q \\ n\leqslant x}} 1. \qquad (209)$$

Consider the contribution to the sum due to those pairs p, q for which $(p-1,q-1) = l < X$. Then, if the congruences

$$\left.\begin{aligned} 2^\nu+b &\equiv 0, \quad \mod p \\ 2^\nu+b &\equiv 0, \quad \mod q \end{aligned}\right\} \qquad (210)$$

are soluble, the solutions form a single residue class, modulo $[p-1, q-1]$. In this case the inner sum is

$$\frac{lx}{(p-1)(q-1)}+O(1),$$

while in the other case it is zero. Hence we need to evaluate the sum of the series

$$\sum_{p,\,q\leqslant X}\frac{1}{(p-1)(q-1)}, \qquad (211)$$

where the summation is over the pairs p, q with $(p-1, q-1) = l$ and with (210) soluble.

Without entering into full details we remark that (210) is soluble if and only if there exists a positive integer $r < l$ such that $2^r b$ is an lth power residue, modulis p and q. Hence, if the condition $(p-1, q-1) = l$ were now replaced by the weaker condition $l\mid(p-1)$, $l\mid(q-1)$, the corresponding sum would be

$$\sum_{r<l} \tau^2(X,l,r), \qquad (212)$$

where
$$\tau(X, l, r) = \sum_{\substack{p \leqslant X \\ p \equiv 1, \bmod l \\ \theta_r}} \frac{1}{p-1},$$

θ_r being the condition that $2^r b$ be an lth power residue, mod p. In like manner the sum (211) itself can be expressed in terms of more recondite sums τ' that are analogous to $\tau(X, l, r)$.

Heuristically $\tau(X, l, r)$ is

$$O\left(\frac{\log \log X}{l \phi(l)}\right),$$

and we therefore expect (212) to be

$$O\left(\frac{(\log \log X)^2}{l \phi^2(l)}\right).$$

Hence summing over l we would foresee that the main contribution to (209) would correspond to small values of l only.

Actually when l is small the methods of Chapter 3 suffice to estimate τ' adequately provided the extended Riemann hypothesis be assumed, the condition $(p-1, q-1) = l$ not causing any essential difficulty. Thus altogether on the Riemann hypothesis there is a satisfactory treatment of what is probably the dominant part of (209).

When l is larger the weaker condition $l|(p-1)$, $l|(q-1)$ may be used instead, since probably only a remainder term is then at issue; similarly the condition that 2 be a primitive root, mod p, may be omitted at the appropriate stage. Notwithstanding, however, the simplifications in the method that are made possible by the introduction of the alternative conditions, the extended Riemann hypothesis is not enough to enable us to deduce an upper bound for $\tau(X, l, r)$ for all the required range of l, and it is necessary to make an additional assumption in order to finish the estimations.

The simplest procedure is to adopt the following additional

HYPOTHESIS A   *Let* $P_b(y; l, r)$ *be the number of primes* p *not exceeding* y *for which* $2^r b$ *is an* lth *power residue, mod* p, *and for which* $l|(p-1)$. *Then, for* $y^{\frac{1}{4}} < l < y$, *we have*

$$P_b(y; l, r) = O\left(\frac{y}{\phi(l) \log^2 (2y/l)}\right).$$

Since the remaining part of the assessment of $\tau(X, l, r)$ can then be completed by partial summation, we are led to the conclusion that

$$\sum_{n \leqslant x} \omega_X^2(2^n + b) \sim C^2 x (\log\log X)^2.$$

Comparing this with (208) in the usual way, we infer that the normal order of $\omega_X(2^n + b)$ for $n \leqslant x$ is $C \log\log X \sim C \log\log\log x$, and hence that the number of such n for which $\omega_X(2^n + b) \leqslant 1$ is $o(x)$. We thus obtain

THEOREM 8 *Let* $|b|$ *be an odd integer exceeding* 1, *and define* $\pi_b(x)$ *to be the number of primes of the form* $2^n + b$ *for* $n \leqslant x$. *Then*

$$\pi_b(x) = o(x)$$

provided both the extended Riemann hypothesis and Hypothesis A hold.

In our presentation of this theme we deliberately avoided any attempt to economize on the hypotheses used. It is, however, plain from Chapter 3 that the Artin conjecture, the extended Riemann hypothesis, and Hypothesis A are all closely inter-related, the earlier sum $P(x, l)$ being the same as the sum $P_2(x; l, 0)$. Since the extended Riemann hypothesis is only used in the treatment of Artin's conjecture in order to achieve satisfactory estimates for $P(x, l)$, we can in fact conclude from the last section of Chapter 3 that a slight variant of Hypothesis A is in itself sufficient to establish the truth of Artin's conjecture and hence of Theorem 8.

A similar but harder problem is posed by the number of representations $\rho(n)$ of large numbers n in the form

$$n = 2^r + p.$$

Erdös [12] in discussing this equation noted in particular that $n = 105$ has a maximal number of representations (i.e. there is a representation for each r such that $2^r < n - 1$) and asked if there were any large numbers n with the same property. Presumably the answer should be in the negative since one would expect that $\rho(n) = o(\log n)$. However, a proof of the latter seems to present extreme difficulty, even the previous method not admitting of an appropriate modification.

It is comparatively easy to shew that numbers n with $\rho(n)$ maximal are relatively scarce. Let $\nu(x)$ be the number of such numbers not exceeding x, it being trivial that

$$\nu(x) = O\left(\frac{x}{\log x}\right).$$

Then, if $\rho(n)$ is maximal, we have for any given s that

$$n-2, n-2^2, \ldots, n-2^s$$

are all prime for $n > 2^s$. Hence, by the upper bound sieve method (note formulae in Chapter 1, §4),

$$\nu(x) < \frac{A(s)\,x}{\log^s x}$$

for any s. This estimation can be considerably improved by letting s be a suitable function of x, although complications are introduced through the need to consider the exponents to which 2 belongs, modulis all the primes (see a recent paper by Vaughan [79]). If Artin's conjecture is assumed, the estimate can be further improved along the same lines, although then it is easier to proceed as follows.

Suppose n is not divisible by at least one prime p less than $\log n$, modulo which 2 is a primitive root. Then, since $p \nmid n$, the congruence

$$n - 2^\nu \equiv 0, \quad \bmod p,$$

is soluble and has a solution ν with $0 \leqslant \nu < \log n$. For such a value ν we have

$$n - 2^\nu > n - 2^{\log n} > \tfrac{1}{2}n > \log n > p \quad (n > n_0)$$

so that $n - 2^\nu$ is not a prime and $\rho(n)$ is not maximal. Hence, if $\rho(n)$ is maximal and if $\tfrac{1}{2}x < n \leqslant x$, then n is divisible by

$$\prod_{p < \log \frac{1}{2}x} p = \exp\left(\sum_{p < \log \frac{1}{2}x} \log p\right)$$

$$> \exp\left\{(C - \tfrac{1}{2}\epsilon)\log \tfrac{1}{2}x\right\} > x^{C-\epsilon} \quad (x > x_0(\epsilon)),$$

where the product and sum are taken only over primes modulis which 2 is a primitive root. Hence

$$\nu(x) - \nu(\tfrac{1}{2}x) = O(x^{1-C+\epsilon}),$$

from which we have

$$\nu(x) = O(x^{1-C+\epsilon})$$

by an easy argument. It is interesting to note that, in Erdös's example of $n = 105$, the number 2 is a primitive root, modulis the two smaller prime factors 3, 5 of 105.

We remark in conclusion that results in the opposite direction are supplied by the now classical work of Romanoff ([65], see also Landau [51]), who shewed by the upper bound sieve method that there is a positive density of numbers n expressible as a sum of a prime and a power of 2.

3   Cullen's primes

The attempt to apply sieve methods to the sequence $2^n + b$ was only attended with conditional success. Nevertheless they can be applied more successfully to other naturally defined sparse sequences such as that of Cullen's numbers $n2^n + 1$, which once had been conjectured to be composite for all $n > 1$ until Robinson shewed that $n = 141$ gave a prime. We sketch below a proof of the new result that almost all the Cullen numbers are composite (i.e. almost all n give rise to composite Cullen numbers).

First we investigate $N_l(x)$, which we define to be the number of integers n not exceeding x for which the corresponding Cullen numbers are divisible by an odd square-free number l. Consider all the solutions of the congruence

$$\nu 2^\nu + 1 \equiv 0, \quad \mathrm{mod}\, l, \tag{213}$$

by taking all those in the first place that satisfy

$$\nu \equiv \alpha, \quad \mathrm{mod}\,\{\phi(l)\},$$

for a given value of α. Then, by Fermat's theorem, all such solutions must satisfy

$$\nu 2^\alpha + 1 \equiv 0, \quad \mathrm{mod}\, l,$$

which determines ν, modulo l, since l is odd. Hence, considering all values of α, mod $\phi(l)$, we see that if $(l, \phi(l)) = 1$ then the solutions of (213) belong precisely to $\phi(l)$ residue classes, mod $l\phi(l)$, and we deduce in this case that

$$N_l(x) = \frac{x}{l} + O\{\phi(l)\} = \frac{x}{l} + O(l). \qquad (214)$$

In order to utilize this formula we require a suitable sequence of primes $p_1, p_2, \ldots, p_r, \ldots$ with the property that $p_i \nmid p_j - 1$ and that

$$\sum_{p_i} \frac{1}{p_i}$$

is divergent. Such a sequence is in fact supplied by the set Q of primes that is defined iteratively by the requirement that $2 \notin Q$ and that $p \in Q$ when $p \geqslant 3$ if and only if $p' \nmid p - 1$ for all $p' \in Q$ with $p' < p$ (or, what is then seen to be equivalent, simply all $p' \in Q$), it being implicit that $3 \in Q$. This sequence, which was first discussed by Golomb [23], was subsequently shewn by Erdös [16] to have the property that

$$\varpi(x) \sim \frac{x}{\log x \log \log x},$$

where $\varpi(x)$ is the number of such primes q not exceeding x. The sequence in particular is therefore such that

$$\sum_q \frac{1}{q}$$

is divergent.

The latter fact, which is all that we need here, is proved below by a method that is simpler than Erdös's because of the weaker nature of the result required. The sieve method used also has an interest of its own in that it involves a method that is intermediate in power between that of the simple asymptotic sieve and that of the more familiar harder sieves.

We suppose to the contrary that

$$\sum_q \frac{1}{q}$$

is convergent. Let $\xi_1 = \log \log x$ and let l_1 be a square-free number composed entirely of prime factors q not exceeding ξ_1. Then $\varpi(x)$ just counts all those odd primes p not exceeding x which are such that $p \not\equiv 1$, $\mod q$, for $q < x$. Hence, by (1), (2), (3), and (4),

$$\varpi(x) \geqslant \sum_{l_1} \mu(l_1) \pi(x; 1, l_1) - \sum_{\xi_1 < q \leqslant \xi_2} \pi(x; 1, q) - \sum_{\xi_2 < q < x} \pi(x; 1, q) - 1$$

$$= \Sigma_1 - \Sigma_2 - \Sigma_3 - 1, \quad \text{say.} \quad (215)$$

In this, by the prime number theorem for arithmetic progressions (Bombieri's theorem is not needed),

$$\Sigma_1 = \sum_{l_1} \mu(l_1) \left\{ \frac{\operatorname{li} x}{\phi(l_1)} + O(x e^{-A_1 \sqrt{(\log x)}}) \right\},$$

since $l_1 \leqslant e^{\vartheta(\xi_1)} < \log^2 x$ for $x > x_0$. Therefore

$$\Sigma_1 = \operatorname{li} x \prod_{q \leqslant \xi_1} \left(1 - \frac{1}{q-1}\right) + O\left(\frac{x}{\log^2 x}\right)$$

$$> A_2 \operatorname{li} x, \quad (216)$$

since

$$\prod_q \left(1 - \frac{1}{q-1}\right)$$

is convergent.

Next, by Lemma 1, we have

$$\Sigma_2 = O\left(\sum_{\xi_1 < q \leqslant \xi_2} \frac{x}{\phi(q) \log (x/q)} \right) = O\left(\frac{x}{\log x} \sum_{q > \xi_1} \frac{1}{q} \right)$$

$$= o\left(\frac{x}{\log x} \right), \quad (217)$$

if $\xi_2 = x^c$ for a suitable constant c such that $\frac{1}{2} < c < 1$.

Moreover, we have

$$\Sigma_3 = \sum_{\substack{p-1=mq \\ \xi_2 < q < x;\ p \leqslant x}} 1,$$

where the variables of summation are p, q, and m. Here

$$m < x/\xi_2 = x^{1-c},$$

and so

$$\Sigma_3 \leqslant \sum_{m < x^{1-c}} \sum_{\substack{p-1=mq \\ p \leqslant x}} 1 \leqslant \sum_{m < x^{1-c}} \sum_{\substack{p-1=mp' \\ p \leqslant x}} 1,$$

in which the last inner sum is

$$O\left(\frac{x}{\phi(m\log^2(x/m))}\right)$$

by an obvious variant of Lemma 2. Hence, since $x/m > x^{\frac{1}{2}}$, we have

$$\Sigma_3 < \frac{A_3 x}{\log^2 x} \sum_{m < x^{1-c}} \frac{1}{\phi(m)} < \frac{A_4(1-c)\,x}{\log x}. \qquad (218)$$

Combining (215), (216), (217), and (218), and taking c to be sufficiently close to 1, we infer that

$$\varpi(x) > \frac{A_5 x}{\log x} \quad (x > x_0).$$

From this we would deduce by partial summation that

$$\sum_{q \leqslant x} \frac{1}{q} > A_6 \log\log x \quad (x > x_0'),$$

which would imply that $\Sigma\dfrac{1}{q}$ was divergent contrary to hypothesis.

Hence $\Sigma\dfrac{1}{q}$ is anyway divergent.

We can now find an upper bound for the number $\kappa(x)$ of Cullen's primes for which n does not exceed x. Letting l_1' denote, generally, a square-free number composed entirely of prime factors q not exceeding ξ, we have by (1) and (214) that

$$\kappa(x) < \xi + \sum_{l_1'} \mu(l_1')\,N_{l_1'}(x)$$

$$= \xi + x\sum_{l_1'} \frac{\mu(l_1')}{l_1'} + O(\sum_{l_1'} l_1')$$

$$= x\prod_{q \leqslant \xi}\left(1 - \frac{1}{q}\right) + o(x) = o(x)$$

by taking $\xi = \xi(x)$ to be a sufficiently slowly increasing function that tends to infinity as $x \to \infty$. This gives

THEOREM 9   *Let $\kappa(x)$ be the number of positive integers n not exceeding x for which $n2^n + 1$ is a prime number. Then we have*

$$\kappa(x) = o(x)$$

as $x \to \infty$.

Similar methods are applicable to other sequences such as $2^n + n^2$. We may therefore conclude that sieve methods can make a modest contribution to our knowledge of the more sparsely distributed sequences, albeit they will not probably lead to the solution of the more interesting problems concerning them.

Bibliography

1. H. Bilharz, Primdivisoren mit vorgegebener Primitivwurzel, *Math. Ann* **114** (1937), 476–92.
2. E. Bombieri, On the large sieve, *Mathematika* **12** (1965), 201–25.
3. B. M. Bredikhin, The dispersion method and binary additive problems, *Russ. Math. Surveys*, **20** (2) (1965), 85–125.
4. V. Brun, Über das Goldbachsche Gesetz und die Anzahl der Primzahlpaare, *Archiv for Math. og Naturvidenskab*, **34** (1915), no. 8.
5. V. Brun, Le crible d'Eratosthène et le théorème de Goldbach, *Videnskaps-selskapets Skrifter, Mat.-naturv. klasse, Kristiania*, 1920, no. 3.
6. J. Chalk and R. A. Smith, On Bombieri's estimate for exponential sums, *Acta Arith.* **18** (1971), 191–212.
7. Chen Jing-run, The lattice points in a circle, *Sci. Sinica*, **12** (1963), 633–49.
8. H. Davenport, *Multiplicative number theory*, Markham Publishing Company Chicago, 1967.
9. R. Dedekind, *Gesammelte Math. Werke*, 1.
10. P. D. T. A. Elliott and H. Halberstam, Some applications of Bombieri's theorem, *Mathematika*, **13** (1966), 196–203.
11. P. Erdős, On the normal number of prime factors of $p-1$ and some related problems concerning Euler's ϕ function, *Quart. J. Math.*, Oxford, **6** (1935), 205–13.
12. P. Erdős, On integers of the form $2^k + p$ and some related problems, *Summa Brasil. Math.* **2** (1950), 113–23.
13. P. Erdős, The sum $\Sigma d\{f(k)\}$, *J. Lond. Math. Soc.* **27** (1952), 7–15.

14. P. Erdös, On the greatest prime factor of $\Pi f(k)$. *J. Lond. Math. Soc.* **27** (1952), 379–84.
15. P. Erdös, Arithmetical properties of polynomials, *J. Lond. Math. Soc.* **28** (1953), 416–25.
16. P. Erdös, On a problem of S. Golomb, *J. Aust. Math. Soc.* **2** (1961), 1–8.
17. P. Erdös, Some recent advances and current problems in number theory. *Lectures on Modern Mathematics* 3, 196–244, New York, 1965.
18. T. Estermann, Einige Sätze über quadratfreie Zahlen, *Math. Ann.* **105** (1931), 653–62.
19. T. Estermann, Proof that every large integer is the sum of two primes and a square, *Proc. Lond. Math. Soc.* (2) **42** (1936), 501–26.
20. P. X. Gallagher, Bombieri's mean value theorem, *Mathematika*, **15** (1968), 1–6.
21. P. X. Gallagher, A larger sieve, *Acta Arith.* **18** (1971), 77–81.
22. M. Goldfeld, A further improvement of the Brun–Titchmarsh theorem, *J. Lond. Math. Soc.* (2) **11** (1975), 434–44.
23. Solomon W. Golomb, Sets of primes with intermediate density, *Math. Scand.* **3** (1956), 264–74.
24. G. R. H. Greaves, Large prime factors of binary forms, *J. Number Theory*, **3** (1971), 35–59.
25. H. Halberstam and H. E. Richert, *Sieve methods*, Academic Press, 1975.
26. G. H. Hardy and Marcel Riesz, *The general theory of Dirichlet's series*, Cambridge University Press, 1952.
27. G. H. Hardy and S. Ramanujan, The normal number of prime factors of a number n, *Quart. J. Math.* **48** (1917), 76–92.
28. G. H. Hardy and J. E. Littlewood, Some problems of partitio numerorum; III: on the expression of a number as a sum of primes, *Acta Math.* **44** (1922), 1–70.
29. G. H. Hardy and E. M. Wright, *Introduction to the theory of numbers*, 4th edn, Oxford University Press, 1964.
30. C. Hooley, An asymptotic formula in the theory of numbers, *Proc. Lond. Math. Soc.* (3) **7** (1957), 396–413.
31. C. Hooley, On the representation of a number as the sum of two squares and a prime, *Acta Math.* **97** (1957), 189–210.
32. C. Hooley, On the representations of a number as the sum of two cubes, *Math. Z.* **82** (1963), 259–66.
33. C. Hooley, On the distribution of the roots of polynomial congruences, *Mathematika*, **11** (1964), 39–49.
34. C. Hooley, On the difference between consecutive numbers prime to n: II, *Publicationes Math. Debrecen*, **12** (1965), 39–49.
35. C. Hooley, On the power free values of polynomials, *Mathematika*, **14** (1967), 21–6.
36. C. Hooley, On Artin's conjecture, *J. reine angew. Math.* **225** (1967), 209–20.
37. C. Hooley, On the greatest prime factor of a quadratic polynomial, *Acta Math.* **117** (1967), 281–99.
37a. C. Hooley, On binary cubic forms, *J. reine angew. Math.* **226** (1967), 30–87.
38. C. Hooley, On the square-free values of cubic polynomials, *J. reine angew. Math.* **229** (1968), 147–54.
39. C. Hooley, On the intervals between numbers that are sums of two squares, *Acta Math.* **127** (1971), 279–97.
40. C. Hooley, On the Brun–Titchmarsh theorem, *J. reine angew. Math.* **225** (1972), 60–79.

41. C. Hooley, On the intervals between consecutive terms of sequences, *Proceedings of Symposia in Pure Mathematics*, American Math. Soc., 1973.

42. C. Hooley, On the greatest prime factor of $p+a$, *Mathematika*, **20** (1973), 135–43.

43. C. Hooley, On the intervals between numbers that are sums of two squares: III, *J. reine angew. Math.* **267** (1974), 207–18.

44. C. Hooley, On the Brun–Titchmarsh theorem: II, *Proc. Lond. Math. Soc.* (3) **30** (1975), 114–28.

45. M. N. Huxley, On the difference between consecutive primes, *Inventiones Math.* **15** (1972), 164–70.

46. J. Ivanov, Über die Primteiler der Zahlen von der Form $A+x^2$, *Bull. Acad. Sci. St. Petersburg*, **3** (1895), 361–7.

47. H. Iwaniec, Primes of the type $\phi(x, y)+c$ where ϕ is a quadratic form, *Acta Arith.* **21** (1972), 203–34.

48. H. D. Kloosterman, On the representations of a number in the form $ax^2+by^2+cz^2+dt^2$, *Acta Math.* **49** (1926), 407–68.

49. E. Landau, *Handbuch der Lehre von der Verteilung der Primzahlen*, Teubner, Leipzig, 1909.

50. E. Landau, *Algebraische Zahlen*, Teubner, Leipzig, 1927.

51. E. Landau, *Über einige nevere Fortschritte der additiven Zahlentheorie*, Cambridge University Press, 1937.

52. Yu. V. Linnik, An asymptotic formula in the Hardy–Littlewood additive problem, *Izv. Akad. Nauk SSSR, Ser. Mat.* **24** (1960), 629–706.

53. Yu. V. Linnik, *The dispersion method in binary additive problems*, American Mathematical Society, Providence, Rhode Island, 1963.

54. A. A. Markov, Über die Primteiler der Zahlen von der Form $1+4x^2$, *Bull. Acad. Sci. St. Petersburg*, **3** (1895), 55–9.

55. J. C. P. Miller, see A. E. Western and J. C. P. Miller, *Tables of Indices and Primitive Roots*, Royal Society Mathematical Tables, vol. 9, p. xxxviii, Cambridge, 1968.

56. B. G. Moĭsezon and M. A. Subhankulov, *Izv. Akad. Nauk Uz. S.S.R. Fiz.-Mat.*, 1960, No. 6, 3–16.

57. H. L. Montgomery, A note on the large sieve, *J. Lond. Math. Soc.* **43** (1968), 93–8.

58. H. L. Montgomery and R. Vaughan, The large sieve, *Mathematika*, **20** (1973), 119–35.

59. Y. Motohashi, A note on the least prime in an arithmetic progression with a prime difference, *Acta Arith.* **17** (1970), 283–5.

60. Y. Motohashi, On some improvements of the Brun–Titchmarsh theorem, *J. Math. Soc. Japan*, **26** (1974), 306–23.

61. T. Nagell, Généralisation d'un theorème de Tchebycheff, *J. Math. Pures Appl.* (8), **4** (1921), 343–56.

62. T. Nagell, *Introduction to number theory*, Almqvist and Wiksell, Stockholm, 1951.

63. K. Ramachandra, A note on numbers with a large prime factor I, *J. Lond. Math. Soc.* (2) **1** (1969), 303–6.

64. G. Ricci, Ricerche aritmetiche sui polinomi, *Rend. Circ. Mat. Palermo*, **57** (1933), 433–75.

65. N. P. Romanoff, Über einige Sätze der additiven Zahlentheorie, *Math. Ann.* **109** (1934), 668–78.

66. H. Salié, Über die Kloostermanschen Summen $S(u, v; q)$, *Math. Z.* **34** (1931), 91–109.

67. W. Schwarz, Über B-Zwillinge II, *Archiv der Math.* **23** (1972), 408–9.

68. A. Selberg, On the normal density of primes in small intervals and the difference between consecutive primes, *Archiv. Math. og Naturvid.* **47** (1943), 87–105.

69. A. Selberg, On an elementary method in the theory of primes, *Norske Vid. Selsk. Forh., Trondhjem,* **19** (1947), 64–7.

70. A. Selberg, On elementary methods in prime number theory and their limitations, *Skand. Mat. Kongr.* **11**, (1949), 13–22.

71. A. Selberg, The general sieve method and its place in prime number theory, *Proc. Int. Cong. Math.* **1** (1950), 286–92.

72. A. Selberg, Sieve methods, *Proc. Sympos. Pure Math.* **20** (1971), 311–51.

73. W. Sierpiński, Sur un problème du calcul des fonctions asymptotiques, *Prace mat.-fis.,* **17** (1906).

74. H. J. S. Smith, *Collected Mathematical Papers,* **1**, Oxford University Press, 1894.

75. G. K. Stanley, On the representation of a number as a sum of squares and primes, *Proc. Lond. Math. Soc.* (2) **29** (1928), 122–44.

76. E. C. Titchmarsh, A divisor problem, *Rend. Circ. Mat. Palermo,* **54** (1930), 414–29.

77. E. C. Titchmarsh, *The theory of the Riemann zeta-function,* Oxford University Press, 1951.

78. S. Uchiyama, On the power-free values of a polynomial, *Tensor (New Series),* **24** (1972), 43–8.

79. R. C. Vaughan, Some applications of Montgomery's sieve, *J. Number Theory,* **5** (1973), 64–79.

80. A. I. Vinogradov, Artin *L*-series and his conjectures, *Proc. Steklov Inst. Math.* **112** (1971), 124–42.

81. A. Weil, On some exponential sums, *Proc. Nat. Acad. Sci. U.S.A.* **34** (1948), 204–7.

82. A. Weil, Sur les courbes algébriques et les variétés qui s'en déduisent, *Actual. Scientif. Ind.* no. 1041 (Paris, 1948).

83. B. M. Wilson, Proofs of some formulae enunciated by Ramanujan, *Proc. Lond. Math. Soc.* (2) **21** (1922), 235–55.